AF540429

FOOD MICROBIOLOGY

Methods of Enumeration

FOOD MICROBIOLOGY

Methods of Enumeration

S.P. Narang
MSc, Microbiologist

A P H PUBLISHING CORPORATION
ANSARI ROAD, DARYA GANJ
NEW DELHI-110 002

Published by
S.B. Nangia
A P H Publishing Corporation
4435-36/7, Ansari Road, Darya Ganj
New Delhi-110002
Ph. 23274050

Email : aphbooks@gmail.com

2021

Typesetting at
Creative Graphics

Printed at :
Balaji Offset
Navin Shahdara,
Delhi-32

Preface

The concept of food safety is different from food purity. Earlier the consumer thought of his food to be free from adulteration. Now a day the concept of food safety is more relevant and important then other factors such as colour flavour etc., unsafe food results in diseases to the consumer. The cause of these diseases is the microbiological contamination of the ingested food(s). This has necessitated ensuring the foods are microbiological safe. This cannot be attained without the proper knowledge of subject of microbiology.

The purpose of this book is to present practical methods for food microbiologists working in food industries, defence food laboratories and other testing laboratories. It also meets the needs of those engaged in teaching food microbiological and plant pathology. Special details have been provided to explain the methods adopted by national or international bodies like AOAC, ISO, BIS, ASC and International Commission for Microbiological specifications for foods.

In some quality insurance laboratories, rapid and instrumental methods are used such as polymerase chain reactions (PCR), Immunomagnetic capture DNA probes and flow cytometry using whole cell Florescence In Situ Hybridization (FISH). These methods have been explained in short form in this book. These are very expensive and beyond approach of middle class industries. The conventional methods of microbiological analysis described in this book are easy to follow, not cumbersome and accurate results are obtained but time consuming.

I am indebted to Scientific community of the world and my friends whose writing, knowledge and experience I could use. My sons Pankaj and Rishi have assisted me in collection of some important information on internet and corrected mathematical calculations. My grandson Raghav and my wife Santosh and my daughter-in-law Ritu have shown lot of patience for not disturbing me while writing. I am thankful to Sh. S.N. Mahindru for encouraging to write this book.

S.P. Narang

Contents

1

Microorganisms in Foods

HISTORY

The use of Microorganisms in Foods product is of old history. Their actual identity established after discovery of Lens which is now a part of Modern Microscope. We now divide the entire world into three kingdoms, Plants, Animals and Microbes or Protista-a-German name. The Microbes are present in innumerable numbers and are of different varieties. They play different of parts in different kind of Food. The presence of Microorganisms in foods is not necessarily an indicator of hazard to the consumer or of inferior quality. Food without presence of Microorganisms is no more a food. The Microbes kingdom includes viruses, bacteria, protozoa, algae, fungi, toadstools, Seaweed etc.

In industrialized countries the percentage of the people becoming ill from food born diseases each year is estimated to be around 30%. Even in the advance country such as in the USA suffer roughly 76 million food born diseases per year, resulting in the hospitalization of 3,25,000 people and 5,000 death (WHO, Food safety and food born illness, 2002). Many of these cases related to the contamination of meat and poultry products.

Cases of Meat/processed food contamination generally in countries like India food born diseases relates to occasions such as social function, where a group of people at one place fell ill from eating the same batch of product. In this situation food safety problem has aroused.

CLASSIFICATION OF MICROORGANISMS

To study scientifically Microbes have been divided into two groups- the lower protists and the higher protists. The lower protists are called prokaryotes examples of which are Bacteria, actinomycetes, Blue- green algae, Spirochetes, rickettsiae bodies etc. The higher protists are called Eucaryotes, example of which are Fungi, Protozoa and Algae. Members belonging to this category posses a normal nucleus surrounded by a membrane. It is believed that higher plants and animals have evolved from the higher protests. Viruses as said earlier, are at the threshold of life as they do not posses a cell and a nucleus viruses are extremely small and can be seen only with the help of an electron microscope.

The word 'VIRUS' is of Latin origin meaning poison. Viruses are not made up of cells as seen in other living organisms. On the contrary the genetic material or the nucleic acid is surrounded by a protein Coat. The Composition of viruses depends upon the type of Nucleic acid present and divided into two groups:

Deoxy virus and Ribo virus. The plant viruses such as TMV always have RNA as the NA. On the other hand, the animal viruses and the bacteria phages (bacterial viruses) are composed of either DNA (Deoxy Ribo Nucleic Acid) or RNA (Ribo Nucleic Acid). However, both types are not found in the same virion.

The viruses can multiply only when they are inside a living cell. Outside a cell they are just like a chemical entity. From this point of view, the viruses cannot be considered as living. The important feature to be noted here is that virus particles can multiply but cannot reproduce their own kind. When introduced into a living cell, the virus takes over the cells, control and directs it to produce its own specific Viral protein and nucleic acid. Thus, new viruses are formed in the host cell. Professor Stanley has pointed out "the moment a virus infects a living cell, it is itself transformed into a living one." A specific virus has a specific and intimate relationship with its host cell and can live only in that particular cell. If it is transferred to another living cell, it loses its capacity to multiply. According to Professor Andrew Lwoff an authority on virus, it is better not to worry whether or not viruses are living, but to accept that viruses are viruses.

The utility of viruses as indicators of food contamination is questionable because of the technical difficulties of demonstrating their presence. Evidence for the occurrence of food-borne outbreak of virus diseases such as infectious hepatitis or poliomyelitis is usually based on after the fact epidemiological investigations rather than Laboratory examinations of the implicated food. Nevertheless, tests for plaque-forming enteroviruses in food have been proposed by Kostenbader and Cliver (1973). Such tests would appear to be most useful in Situations where virus contaminants persist even though tests for the common bacterial indicators give negative results.

Viruses are more resistant to gamma radiation than are other microorganisms. Considering the high radiation doses necessary to inactivate viruses, the use of radiation to inactivate viruses is not practical. Thermal inactivation seems to be an acceptable.

MICROBIAL EFFECT ON FOOD

Microorganisms reduce man's food supply by creating an unpleasant smell and discolored appearance, such food is either to be destroyed or discarded and unbearable losses to the nation. When high numbers of microorganisms are present, it contains microbial toxin and it is advisable to destroy the food.

SPOIL OF FOOD BY MICROORGANISMS

Food born illness is common in most of the countries. Stale Meat & Meat Products, Cooked Rice, Milk & Milk products, Potatoes, Drinking Water, Soft Drink and Canned Food may even can cause death to a man/ animal / birds when consumed. They play a friendly part as well as enemy part. It is difficult to place an organism in one category since it may have different function in different foods. An organism may helpful or inert in one food but spoils in other food due to the different composition of the foods.

FOOD BACTERIA

Very few species of bacteria are present in food as compared to total number of species present in nature. Here we are concerned with food bacteria. According to Bergery's manual bacteria are grouped according to Gram's reaction, morphology and relation to growth with bacterial cell structure or without oxygen. Here we will discuss bacteria found in food on same line.

With the help of conventional light microscope it is possible to know only the outline of bacterial cells. Detailed knowledge about bacterial cytology or fine structure of bacterial cell is now possible with electron microscope.

BACTERIAL CELL STRUCTURE

Diagram of Cell

There are three shapes of cell among the true bacteria: Rod, Sphere and Helix. These cells may occur singly or in various groupings.

Bacterial Cell Structure

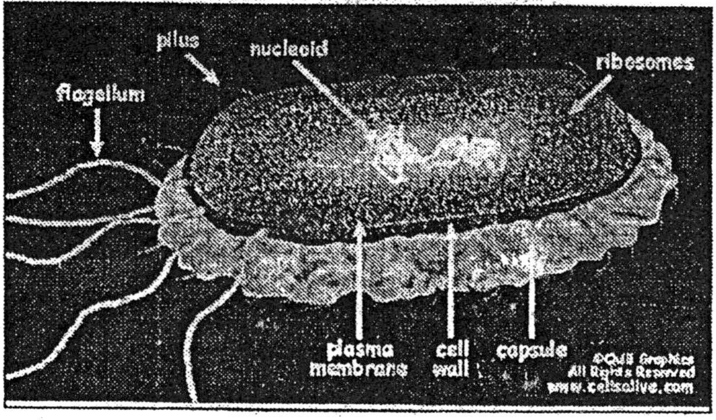

- Nucleoid
- DNA
- Ribosomes
- Cell wall
- Plasma membrane
- Outer membrane
- Capsule
- Elagella
- Pilus

CELLULAR GROUPINGS

Various kind of cellular groupings are seen in bacteria. The rod shape is probably most common form and also occurs in chains called streptobacillus as they occur in anthrax disease. The spherical bacteria or the "COCCI" often occur in chains or in various groupings. The plane of division decides the type of clustering: the division may be in (i) one plane (ii) more than one plane resulting in five types of groupings.

The division in one plane may result in daughter cell pairs remaining close together which are called diplococci as in "pneumonia bacteria" Diplococcus pneumonia. The cells may occur in chains and the individual Cocci may remain near one another as in "Streptococci". An example is seen in streptococcus Salivarius causing severe inflammation of throat. These are present in spices also.

(1) In Gaffya tetragena, the division occurs in two planes at right angles to each other. This results in tetrads or plates of cells which are seen in tuberculous sputum, pus of abscesses and in spinal fluid in case of meningitis.

(2) Division in three planes at right angles to each other results in cubical packets as in sarcinabutea a saprophyte in water, soil and air. But if the plane of division is at various angles, it results in irregular bunches resembling clusters of grapes. An example is seen in staphylococcus aureous which causes boils, abscesses etc.

The Capsule: Many bacteria have a shiny covering which closely adhere on each cell as a distinct layer. The gelatinous covering is known as capsule. In some species production of capsule is a critical feature which determines ability of some pathogens to initiate infection in host tissues.

LABORATORY METHOD OF IDENTIFICATION OF BACTERIAL CELL

The Gram reaction: It is a sample staining technique of bacteria which is named after its discoverer Christian Gram, a Danish physician. The bacteria readily react to this stain which is one of their most important cytological features. In this procedure, the bacterial cells are stained with the crystal violet dye which stains all the bacteria blue. Next the stained bacteria are treated with Iodine solution and then they are washed with alcohol. While the Gram positive bacteria retain the crystal violet colour, the Gram negative bacteria are decolorized. This is due to fundamental difference in the structure of the cell wall of Gram positive and Gram negative bacteria. The division of the bacteria into Gram positive and Gram negative forms correlates very well with their other characteristics. For example, all spore forming bacteria are

Gram positive, all polarity flagellated ones are Gram negative. The Gram-negative bacteria are generally more resistant to many antibiotics, safranine may be used as a counter stained, to stain the Gram negative bacteria an orange red. The staining procedure is as under:

Chemicals:

Crystal violet Solution	
Crystal violet (85-90% dye content)	2g
Ethyl alcohol (95%)	20ml
Ammonium oxalate	0.8g
Distilled water	80ml

Directions: Dissolve crystal violet in alcohol and ammonium oxalate in distilled water. Mix two solutions and store this mixture for 24 hours before use:

Iodine Solution

Iodine	1g
Potassium iodide	2g
Distilled water	100ml

Direction: Grind potassium iodide and iodine together in a mortar, adding small increments of water while grinding. Rinse resulting solution into a volumetric flask and bring the volume to 100ml.

Counter stain

Formula	Safranin 0	0.25g
	Ethyl alcohol	100ml
	Distilled water	100ml

Directions: Dissolve safranin in ethyl alcohol and mix resultant solution with distilled water.

Staining Procedure:

1 Take young culture (18-24 h) on a clean slide, spread it by pouring some water and finally fix it by moving the slide in flame.
2. Stain with Crystal violet solution for 2 minutes. Take care that entire smear is covered with liquid at this and at all subsequent stages.

3. Holding slide at an angle of 45°, wash crystal violet solution off smear. When all traces of crystal violet have been removed, allow iodine solution to act for 1 minute.
4. Pour off iodine blot dry and holding slide at an angle of 45°, wash this slide with 95% ethanol (or industrial methylated spirits) until no more violet stain runs from slide (only 5-15 s)
5. Rinse under tap and stain with an aqueous solution of safranine 0 for 15 s.
6. Wash slide well and blot dry; examine Gram-positive organisms stain blue, Gram-negative organisms stain red.

STAINING OF BACTERIAL FLAGELLA

Diameter of a bacterial flagellum is below the limit of resolution of light microscope and for this reason flagella are normally not visible. By use of special staining techniques an appropriate stain can be made to buildup around each flagellum, thus increasing its apparent diameter. This enables flagellum to be visualized under light microscope and arrangement of flagella on bacterial cell to be determined. Specifically, flagella stain is of use in differentiating members of psedomonadaceae, which have polar flagella, from members of the Enterobaceriaceae, which have peritrichous flagella (when motile).

Good results in flagella staining are obtained only with difficulty, as the extremely delicate flagella easily become detached from the bacterium.

Most commonly used stain is silver salt to deposit silver on flagella and secondly those in which basic fuschin are deposited on flagella.

The silver deposition method described below is a modified Fontana method (Rhodes 1958)

Procedure

Only slides that are completely clean should be used. The bacteria should be grown on agar slopes at 3° – 4°C below optimum growth temperature, the slopes having first been moistened with one or two drops of sterile distilled water.

1. Flame a microscope slide in a spirit lamp for about 5s. Place on a staining rack to cooling and then draw a halfway division with a wax pencil.
2. With a pipette add 2ml of sterile distilled water to a young actively growing, slope culture usually about 18 hrs old and gently suspend the growth by careful agitation and rotation of the test tube. Do not use wire loop. Transfer to a clean test tube, check for motility with a hanging drop preparation, and dilute the suspension with distilled water until only slightly turbid. Place in an incubator at 20°-30° C for 30 minutes and then remove a large loopful of the suspension to one end of the cool microscope slide. Tilt the slide until the drop runs to the central pencil line. Dry in air at room temperature. Do not heat fix the film.
3. Cover with hot Fontana silver solution and stain for 5 minutes, rinse gentle but thoroughly with distilled water to remove all traces of mordant.
4. Renewing the stain once a minute. Do not allow staining solution to get it dried from any portion of the smear-keep this covered with liquid at all times.
5. Wash with water, allow to dry in air and examine

THE STAINING OF BACTERIAL SPORES

Bacteria in the genera bacillus and clostridium produce endospores, which are highly resistant to high temperature, lack of moisture and toxic chemicals. The endospores are also resistant to bacteriological stains and in a smear stained by Gram's method, they can be seen as colourless areas in the vegetative organism which stain Gram's positive. However, once stained the spores tend to resist decolorization.

Procedure for Bartholomew and Mittwer spore staining method

1. Prepare a smear in the usual way, but heat fix very thoroughly by passing thorough a
bunsen flame 20 times.
2. Stain for 15 minutes with a saturated aqueous solution of malachite green.
3. W[illegible]h [illegible]ntly with cold water for 10s.

4. Counterstain with a 0.25% solution of safranine for 15s.
5. Wash with water and blot dry,
6. Examine under oil immersion objective

THE DEMONSTRATION OF BACTERIAL CAPSULES PROCEDURE FOR THE WET INDIAN INK FILM

1. Place a loopful of Indian Ink on a very clean microscope slide.
2. Mix into Indian Ink a little of bacterial culture or suspension.
3. Place a cover slip on mixture avoiding air bubbles and press firmly with blotting paper until the film of liquid is very thin.
4. Examine with the high power dry objective or the oil-immersion objective. The capsule will be seen as a clear area around the bacterium.

Note: A control film of ink always be prepared as Indian Ink may occasionally become contaminated with capsule bacteria.

Negative Staining

This is a very simple and effective method for demonstrating the external shape of bacteria in a smear preparation. The bacteria are surrounded by a thin films of black dye and appear as white objects upon a grey background.

Procedure

1. Prepare a very thin smear in the usual way, using a clean, grease free slide.
2. At one end of the slide place one drop of negrosin solution (2%)
3. Take another microscope slide. Lay of one end on the first slide at an angle of 30°C touching the drop of negrosin and use it to push negrosin across the surface of first slide. The smear will thus be covered with a thin, even, film of dye.
4. Allow the dye to dry and examine the slide under oil- immersion objective.

Ziehl-Nelson Method for staining acid-fast bacteria

Members of the Genera Mycobacterium and Nocardia can be differentiated from many other organisms by this staining technique. Nocardia is present in fish, small animals and sputum's of tuberculosis patients. The genus has been named in honour of medical scientist Nocard.

The Ziehl : Neelsen method consists firstly of staining the organism with a hot, concentrated dye. Once stained, the cells resist decolorization with acid, they are thus acid fast. Decolorization is effected with suitably strong acid 20% of sulphuric acid and the smear is then counterstained with methylene blue solutions. Acid-fast-bacteria stain red, other bacteria and the background stain blue. When 1-5% sulphuric acid (or hydrochloric acid) is used instead of 20% sulphuric acid, the technique is used for identifying saprophytic mycobacteria, nocardiae and certain coryneform organisms.

Procedure

1) Cover the slide with strong Ziehl Neelsen's carbol fuchin and heat underside of slide with a lighted alcohol-soaked swab. Stop heating when the slide steams. Keep slide hot and replenish the stain if necessary, taking care not to allow the smear to become dry. Heat for 5 minutes, not allowing staining solution to boil.
2) Wash well.
3) Decolorize with acid alcohol (1:5) or with 20% sulphuric acid. The excess stain is removed as brownish solution, and smear will become brown. Rinse in water, when the film will appear pink once more. Apply more acid and repeat the rinsing several times until the film appears faintly pink upon washing.
4) Wash Well.
5) Counter stain with locffer's methylene blue for 5 minutes.
6) Wash well and carefully remove the stain deposits from back of slide with filter paper. Blot dry and examine.

Examination of cultures for Motility by "Hanging Drop Preparations"

The culture under test should be Broth culture 18-24 hr. old, or a small amount of culture from an 18-24 hr agar slope can be emulsified gently in a drop of broth or normal saline, taking care that the emulsion is not too dense. Using following procedure, a drop of culture is suspended from a coverslip over the depression in a hollow-ground slide.

Procedure

1. First place a little immersion oil round the edge of the depression in the slide. Then with a wire loop, transfer a small loopful of the culture to a clean dry coverslip laid on the bench. Do not spread the drop.
2. Invert the cavity slide over the coverslip so that the drop is in the centre of the cavity and press, the slide down gently but firmly so that the oil seals the coverslip in position. Invert the slide quickly and smoothly and the drop of culture should now be in the form of a hanging drop. The preparation should be examined without delay and as quickly as possible.
3. When examining hanging drop preparations the sub stage condenser of the microscope should be reached down slightly from the normal position and the iris diaphragm should be almost completely closed. Not only does excessive illumination render the unstained organism invisible, but also the heating effect may cause them to lose their motility.
4. First, use low-power objective to focus on the edge of the drop, moving the slide until the edge of the drop appears across the centre of the field. This should be recognized easily as minute droplets of condensed water can usually be seen on the other side of the line which represents the edge of the hanging drop. Then place the high power (x40) dry objective in position and refocus the edge of the drop. If necessary, open the iris diaphragm slightly. The bacteria should now be seen easily, particularly towards the edge when reduction in the depth of liquid assists in keeping the organism within the depth of field. The oil immersion objective should not be used because the focusing movements of the objective would be mechanically transmitted to the coverslip causing streaming in the culture liquid which would seriously hinder observation and may even be misinterpreted by an untrained observer as motility of the organisms.

It is necessary to distinguish between Brownian movement or drift in one direction caused by the slide being slightly titled and true motility.

The characteristics of motility are used for identification e.g. Listeria displays a tumbling motion, cytophaga a gliding motion. From known stock culture one can identify the organism.

Examination of Microbial Colonies

It is sometime useful to examine microbial colonies microscopically. This is especially valuable in the case of microfungi. Many stereomicroscopes have an adjustable incident light source built into the nose piece, with a choice of objective (x-10, x-20 and even x-40)

Fluorescence Microscopy

Fluorescence is a phenomenon in which light energy at one wavelength is absorbed by a substance and immediately-re-emitted as light of a longer wavelength. A substance showing such a re-emission is known as fluorophore. Some naturally occurring biological substances (e.g. chlorophyll) display fluorescence, this is called auto fluorescence. In Microbiological microscopy specimens are usually stained with fluorescent dyes like accridine orange. It acts as a metachromatic stain, because whereas the monomer fluoresces green (around 525nm) linked dye molecules fluoresce red (around 650 nm). The dye- interactions resulting in red fluorescence can occur when the dye attaches to RNA, whereas when the dye attaches to DNA such dye-dye interactions are absent and the resulting fluorescence is green. The light source for a fluorescence microscope can be halogen-quartz lamp, a high pressure mercury vapour lamp or a xenon arc. The fluorochromes can be conjugated with antibodies to give immunofluorescent stains which permit the detection of specific organisms. Such a technique could be employed, for example, to study the location of specific pathogenic organism within the host tissues.

Transmitted UV-light-Fluorescence Microscopy

Because glass absorbs UV light, the condenser lens system should be made of quartz. The objective and eyepiece lenses can be made of glass as usual, as it is visible light re-emitted by the specimen that will be observed. However, there needs to be steep cut UV filter in the light path to protect the eyes, and these should also be a UV filter screen in front of the microscope stage to control scattered UV light.

Direct Epiflorescence Microscopy

In this method the existing light is directed at the specimen from above through the objective lens system. The light is so designed that focusing the objective on the specimen also focuses the existing light.

The method of microscopy finds particular application in combination with membrane filtration. The direct epifluorescence filter technique was originally developed for the rapid counting of bacteria in raw milk and in heat treated milk. The maximum sensitivity of the method is around 10^3-10^4 bacteria per gram.

Membrane filtration is most readily applied to samples that are liquid and which by appropriate treatment can be made to pass through the membrane filter e.g. milk is treated with protolytic enzyme to break down bovine somatic cells, and with a surfactant to disperse fat globules.

2

Growth of Microorganisms

Bacterial study and identification depend on the type of Bacteria and media used for cultivation. The Microbiological quality of food depends on detection of viable organism, type of material suitable for growth. Netruit materials provided in a form suitable for growth are known as culture media and the growth itself is a culture. Commonly used media are given in the chapter 13.

The culture method helps to identify microbes and also to find organisms in test material of low microbial content. When such material is inoculated on to a suitable culture medium, each organism present is the material multiplies many times giving rise to a colony of bacteria. For satisfactory growth of bacteria on artificial media, the proper temperature, right amount of dilution and pH must be provided. The culture medium itself must contain the necessary nutrients and growth promoting substances and must be free from contaminating microorganisms i.e. it must be sterile.

Types of Culture: Culture media are of two types: Liquid and Solid

It is often desirable that the culture medium be a solid one since microbes can be visualized on surface Agar (Japanese Seaweed) is a solidifying agent that is widely used. It melts completely at the temperature of boiling water and solidifies when cooled to about 45°C. It has no effect on bacteria. It does not provide any nutrient to bacteria. Numerous enriched materials are added in different culture media: Carbohydrate, serum, bile salt etc. Carbohydrates are added to increase the nutritive value of the medium and to indicate the fermentation reaction of the microbes being studied. Serum, blood and ascetic fluid are added to promote the growth of fastidious organisms like meningococci etc. Dyes are added to the culture media to act as indicators to detect the formation of acid when fermentation reaction occurs or to act inhibitors of growth of certain bacteria but not of others. An example of an indicator dye is phenol red, which is red in an alkaline or neutral medium but yellow in an acid one. An example of an inhibitory dye is gentian violet which inhabits the growth of most Gram +ive bacteria.

In actual practice solid media are available which may be reconstituted in distilled water.

Culture Techniques: Transfer of microorganism is normally carried out with platinum or nicrohme wire of a fixed diameter and a loop is made about 4mm. These wires are sterilized by flame at every stage before using in next step.

Cotton wool plugs removed from test tubes before subculture should be held at the projecting surface only, the inner surface must be protected from contamination and may be lightly flamed before replacement. The mouth of tube or flask should be flamed momentarily after removing and again immediately before replacing, the plug. Once the cotton wool plug is removed, exposed tubes are liable to atmosphere contamination. This may be minimized by holding the tubes in an inclined position near the flame and by not delaying the replacement of the plug.

INCUBATION OF CULTURE

The inoculated media should be incubated at the optimum temperature of the organism concerned except in certain case of certain special media (e.g. gelatin stabs are usually incubated at 22°C). Cultures to be examined for motility of flagella type should be incubated at temperature 3 – 5°C below that which is optimal for growth. In studies of spoilage potential, the incubation temperature is determined by the predicted or recommended storage conditions for the food in question, routinely 24hrs, in an incubator at 37°C temperature. Plate cultures should normally be incubated in the inverted position to prevent condensed moisture from falling on to the bacterial growth.

Microbes when, thus incubated multiply rapidly and within a few hours. There are many more organisms in the culture which appear in the form of colony. The colony has characteristic such as size, shape, colour and texture. These are fairly consistent for each species and are valuable in differentiating one species from another. A specimen may contain more than one type of bacteria. In such a case, there is need to purify the culture. Purification can be done by picking up a colony and plating it on to another culture medium. This will allow the growth of one particular bacteria to be studied.

These bacteria can be identified by their colony characteristics and different biochemical tests.

MAINTENANCE OF PURE CULTURE IN THE LABORATORY

The most suitable method for the maintenance of pure culture depends on the characteristics of the particular organisms and the method used must be chosen accordingly, the most convenient for a routine laboratory is a freezer operating from –60°C to –85°C.

Agar Slopes Cultures

Many organisms are grown on the surface of agar slopes (e.g. Nutrient agar, malt Agar) the choice of medium depends on the organism). Slopes are made in the tubes or screw capped bottles, the latter being preferable as there is less risk of drying. The grown culture may be stored in the dark at room temperature or in the refrigerator and these are subcultured at the interval of between 1 month and 2 years depending on the species.

Maintenance of Lactic-Acid Bacteria

Lactic acid bacteria will not grow well on the surface of solid media incubated aerobically, and are more suitably maintained in liquid medium e.g., yeast glucose chalk litmus milk or Robertson's cooked meat medium being subcultured at intervals of 2-4 months. Media containing Milk or added sugars should also contain chalk to buffer against the development of too low a pH. After bacteria have died in the coagulated milk layer in yeast glucose chalk litmus milk, viable organisms may still remain associated with the chalk sediment.

Maintenance of Cultures of Aerobic Bacteria

Anaerobic bacteria will not grow on the surface of solid media incubated aerobically and best kept in a medium providing reducing conditions (e.g. Robertson's cooked meat medium) and subcultured at intervals of up to 1 year.

Preservation under Oil

Cultures are first grown on agar slopes as described under agar slopes culture and then completely covered with sterilize liquid paraffin or mineral oil. (To sterilize liquid paraffin, dispense flask in shallow layers and sterilize in a hot air oven at 160°C for 1-2h).Culture maintained in this way will generally remain viable for seven years without subculturing. Microfungi cultures are maintained in this way.

Freeze-Dried Cultures

In this process cultures are freeze-dried or lyophilized and stored in sealed glass ampoules under vaccuum. Ampoules can be stored at room temperature or in a refrigerator. Culture preserved in this way remains viable for over several years. This method is being followed by culture collection centre and dispatches are made to outstations.

Desiccated Serum Suspensions

Those laboratories not having freeze drying machine can follow the procedure described by Alton and Jones (1963). Suspend a loopful of growth from a 24-48 h culture in 2ml of sterile serum and place one drop into each of a number of sterile plugged 50x6mm tubes. Dry in an evacuated desiccator over phosphoric oxide for 7 days at 5°C, re-evacuating once after 24hrs. Place each tube (containing a thoroughly dry suspension) in a large soda glass test tube, evacuate and seal in a Bunsen flame. Store the culture in a refrigerator in a deep freeze. Such cultures may remain viable for years.

PLATE CULTURES

To know the amount of organisms, a large surface area of medium is required to grow. The Agar medium is allowed to solidify as a thin layer in a petridish. About 10 to 15 ml of melted sterile medium is poured into a sterile petridish, care being taken to avoid contamination. The mouth of flask containing medium should be flamed after removal of cotton wool plug and lid of petridish to be opened near flame.

The poured plates are allowed to cool as to solidify the medium and dried. The presence of moisture on surface of medium will interfere with production of discrete colonies. Plates are dried in an incubator at 37°C for 30 min, to 1 hr. Lid of petridish is first laid in incubator and part of the dish containing medium is inverted and placed with one edge resting on lid. This method of drying helps to avoid contamination from dust.

Streaks Plates-method of Separation of mixed culture

A sterile platinum wire loop is charged with bacterial load and streaking is done as mentioned below.

1. The inoculum is spread evenly over a small area towards edge of plate.
2. The wire loop is sterilized and used to make strokes as shown in Figure below.

For characterization and identification of bacterial cultures, it is possible to use a standard form of description chart that contains a series of descriptive terms to be underlined as appropriate and a number of blanks to be filled in against the various characteristics and tests. A number of bacterial genera or even species are identifiable by eye with experience on basis of colonial morphology, pigmentation, etc. combined with a knowledge of the source of culture e.g. Bacillus cereus var. mycoides, serratia marcescens. Fluoresces, pseudomonas and proteus (when swarming) are quickly recognized.

MORPHOLOGICAL CHARACTERS

A. Gram reaction
B. Shape, size and arrangement of organisms
C. Motility
D. Presence of endospores, capsules and flagella
E. Reaction of Ziehl-Neelsen and any other special stains

Surface Colonies on Solid Media

1. Shape- Circular, irregular, rhizoid

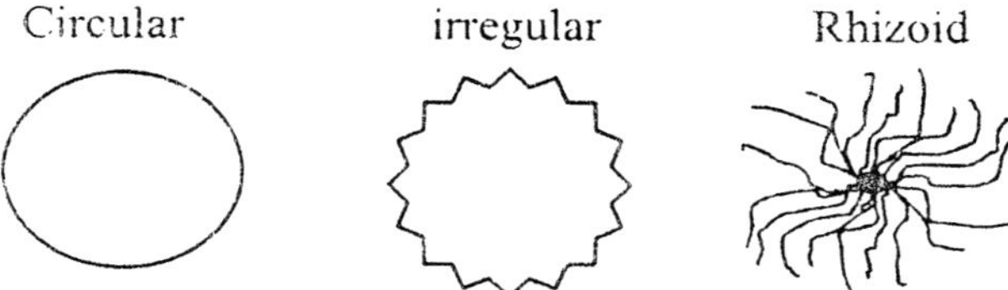

2. Size- Record diameter in millimetre, pinpoint less than 1mm in diameter.
3. Colour of colonies, soluble or insoluble in medium.
4. Opacity-Transparent, translucent, opaque.
5. Elevation- Flat, raised, convex, umbonate

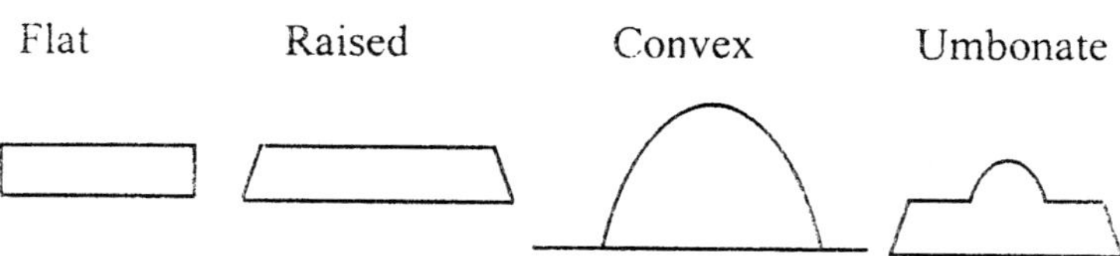

6. Surface Smooth, rough, dull, glistering
7. Edge, entire, undulate, globate, dentate, rhizoid

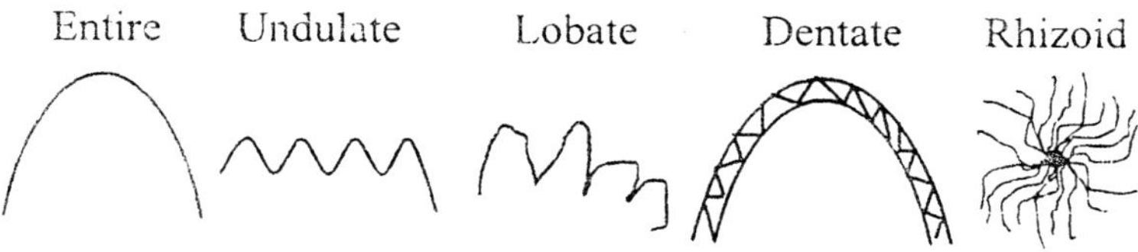

8. Consistency (tested by touching with a sterile wire loop): butyrous, viscid, granular

9. Emulsifiability: Easy or difficult in water, forms uniform turbid suspension, forms granular suspensions, does not emulsify

BROTH CULTURE

A) Amount of growth: None, scanty, moderate.
B) Surface growth: profuse present or absent, formation of ring, pellicle that disintegrates or not on shaking.
C) Turbidity: uniform, flocculent or absent.
D) Deposit-Amount: granular, flocculent, viscid, disintegrates or not on shaking

Shake Culture

Growth: on surface, in depth of tube (record depth to which growth occurs, bearing in mind that insufficient mixing of the inoculum may have stopped the organisms reaching the bottom), position of optimal growth.

THE MICROFLORA OF FOODS

Microorganisms in wide variety generally contaminate foods on farm and during transport to processing plant and also on storage. Some organisms grow, survive or die depends on nature of food, environment and processing procedure. Since these factors are unique for each type of food, microorganisms present on each food are likewise unique to that food. This knowledge about microorganisms represents the science of food microbiology.

A food is spoiled when bacteria, yeast or molds have changed odor, flavour, appearance or texture to an unacceptable degree. Soured fresh milk is one example. However, souring in cheese, buttermilk is acceptable and the particular organism is not considered a spoiling agent. However, this may be classified only in particular foods. A major aim of processing is to make a perishable food shelf stable but these are not always sterile and harbor a selected flora. An adequately long shelf life is a primary goal of good manufacturing practice and the science of food microbiology.

In rapidly perishable foods e.g. those with high moisture content and neutral pH-bacteria grow more rapidly than yeasts/

molds and therefore spoil the food. In food less favourable to microbial growth, bacteria fail to grow or grow very slowly and the yeasts or molds predominate. Thus at temperature between – 5°C and –10°C where fluid water is limited molds predominate. Likewise in foods whose pH is below 5 or whose aw is below 0.90, yeast or molds may grow. The microbial growth in such foods could have mycotoxin but not a bacterial toxin.

The fate of pathogenic and toxigenic organisms in the contaminating flora of a perishable food depends upon both the physical and biochemical environment and the nature of the competing flora- the biological environment staphylococci in small numbers, are relatively frequent contaminations of foods but they compete poorly with the spoilage of flora of most foods of high aw. However, in foods which contain concentrations of salts or sugar that inhibit development of the competing microflora; the number of staphylococci of food is a special concern about public health point of view.

Spoilage of food is a special concern about public health point of view.

Certain intestinal pathogens such as salmonella, shigella and cholera vibrissa are not acceptable even in small quantity in ready to eat food; other pathogens cause illness only if they reproduce to very high levels in the food. This is true of staphylococcus aureous, Escherichia Coli, Clostridium perfringin, Vibrio parahaemolyticus and Bacillu scereus. The organism of this group are harmless in small number, unavoidably present is many raw foods. The normal spoilage flora frequently inhabits growth of pathogens. For this reason, one should not reduce the spoilage flora to a point where pathogens can grow readily, for some foods might become a health hazard before they spoil.

Some food-spoilage bacteria contribute importantly to the human diet in fermentation and similar processes.

In canned food some spores are not destroyed by heat especially in canned meat and later on germinate and grow into active vegetative cells (Kenyam et al. 1961). The activation temperature upon the species i.e. incubator temperature for clostridium botulism is 65°C to 85°C for mesophilic spore forming bacteria and 100°C for spore forming thermopiles.

METHODS FOR MICROBIOLOGICAL EXAMINATION

Intrduction: Methods outlined below are widely used and generally accepted internationally as well as in India. These are followed by Bureau of Indian Standards, Agmark, Prevention of Food Adulteration Act (PFA), Food and Drug Administration (FDA), AOAC (Official methods of Analysis of the Association of official Analytical Chemists) and ISO standards. These methods should provide basis for legal judgement on microbiological quality and safety of food, particularly food moving to international market.

Food sampling procedures should be designed statistically, only then can the results of microbiological analysis provide a basis for statistically valid conclusions about microbiological quality of batches of foods from which samples were drawn.

Preparation and Dilution of Food Sample

For isolation and enumeration of microorganisms in all non-liquid foods are treated with diluents to release those microorganism which may be embedded within the food or within dried or gelatinous surface film.

Most common practice is to use an electric driven mixing device called blender. A homogeneous suspension of food and micro-organism is thus prepared which permits the preparation of dilution for use in various enumerative procedures. Standardization of this preliminary procedure is important. Prolonged use of mixer and high speed may cause injury to microbial cells, either mechanically or by the heat generated. This excessive mixing may result in reduced count. Insufficient mixing in contrast may not release embedded bacteria or provide a heterogeneous distribution within the suspension. It requires lot of experience to control the speed of blender and mixing time.

Recently a new mixing procedure known as stomaching has been developed which is satisfactory for foods. In this technique the food sample and diluents are put into a sterile plastic bag which is than vigorously pounded on its outer surfaces by paddles inside a machine called a 'Stomacher'. After samples are removed for analysis, the bag and its remaining contents can be discarded and the machine is immediately ready for use.

The stomacher is not efficient as blender for breaking up fat globules. Food with fat more than 20% show lower result with stomacher. Addition of 10% Tween 80 or other non toxin surfactant will overcome the problem (Sharpic and Harshaman 1976).

For accurate enumeration of living organisms the nature of diluent used is very important. Many commonly used diluents are distilled water, saline solution, ringer solution and peptone water. Some are toxic to certain organisms and particularly if the time of contact is prolonged (King and Hurst 1963).

The presence of food material may protect many organisms from toxin effects. A general purpose dilution medium used is 1.50% peptone water. Physiological saline with 0.10% peptone added proved to be the most reliable. Peptone (0.10%) in saline (0.85% NaCl) is recommended by international standard organization in its proposed procedure for aerobic counts at 30°C. The diluents are to be used in a defined composition and period of exposure. For isolation and enumeration of particular species or groups known to be sensitive to diluents containing inorganic salts or to distilled water, the least toxin diluent is to be preferred.

Particularly 0.10% Peptone in distilled water or in saline best meets this need. The pH of diluent must be 7.

Preparaing the Dilutions

1. Holding a sterile 1ml or 1.1ml pipette vertically in the sample dipping not more than 3cms below the surface of sample and suck with sterilize pump. Remove excess liquid adhering to outside of pipette. Transfer pipette to first tube of dilution series with the tip touching side of tube 2-3 cm above the level of diluent. Pipette must not connect diluting fluid. Press the pump to release the sample in diluent. Discard this pipette into a used-pipette jar containing a suitable disinfectant, and label the first dilution tube $\frac{1}{10}$ or 10^{-1}.
2. Using a fresh sterilize pipette, mix the contents of first dilution tube sucking up and down to 1ml mark 10 times (or by rotating the tube between hands). Then withdraw 1ml of this first dilution and transfer to second tube of sterile diluent. Pump out the contents of pipette as stated above. Discard this pipette and label the second dilution tube $\frac{1}{100}$ or 10^{-2}.

3. Take a fresh sterile pipette and a further dilution tube; prepare in the same way, a 1 in 1000 or 10^{-3} dilution.
4. Further dilution of 10^{-4}, 10^{-5} etc. can be made similarly as required depending on the probable bacterial content of the sample. Spoil meat and river water can give count more than 10^{-12} per ml.

Preparing the Plates

1. Take a fresh sterilized pipette of 1ml capacity; mix the contents of the final dilution tube e.g. $\frac{1}{1000}$ by sucking up and down 10 times. Withdraw 1ml of the dilution and transfer the contents of a sterile petridish. Allow 3 seconds to elapse and then touch the tip of the pipette against the dish away from the previous inoculum and gently blow out the remaining drops.
2. The same pipette can be used to transfer 1ml from $\frac{1}{100}$ the dilution to a sterile petridish, but before taking the sample raise and lower the $\frac{1}{100}$ dilution in the pipette 3 times to rinse the sides of the pipette and also to give the dilution a final mixing.
3. In the same way the above used pipette can be used to transfer into petridish 1ml of the $\frac{1}{10}$ dilution and then of the original sample.

Pouring the Plates

1. To each plate, add 10 to15ml of molten agar medium and immediately mix the medium and inoculum. Fewer than 20minutes and preferably fewer than 10 minutes should elapse between making the dilution and pouring the agar.
2. Immediately mix the medium with the agar medium by tilting and rotating the petri dishes. A satisfactory sequence of steps is as follows:
   (a) Tilt dish to and fro 5 times in one direction.
   (b) Rotate it clockwise 5 times
   (c) Tilt it to and fro again 5 times in a direction at right angles to that used the first time
   (d) Rotate it counterclockwise 5 times.
3. As a sterility check, pour one or two plate with un- inoculated agar medium and with an un-inoculated diluent. After incubation the number of colonies on these plates should not be enough to distort the count by more than one unit in the final count.

4. Allow the plates to set, then invert and incubate at appropriate temperature, it is essential to label the plates correctly.

Counting the Plates

Select two plates corresponding to one dilution and showing between 30 and 300 colonies per plate. Count all colonies on each plate, using the colonies counter and tally register. Take arithmetic average of two counts when replicate plates have been prepared at each dilution.

Calculate the microbial concentration in the original sample and report the count N of c.f.u. per gm or per ml of sample by dividing the dilution d, (e.g.10^{-2}) used

$$N = \frac{n\bar{x}}{nd}$$

In ISO standard methods another procedure has been adopted for estimating the colony count:

$$N = \frac{\sum C}{(n_1 + 0.1n_2)}$$

where C is the sum of all colonies counted on all dishes containing at least 15 colonies but not more than 300 colonies, n_1 the number of dishes retained in the first dilution, n_2 is the number of dishes retained in the second dilution, and d is the dilution (e.g.10^{-2}) corresponding to the first dilution counted.

In ISO standards an example is given where, at the first dilution retained (10^{-2}), two dishes contained 168 and 215 colonies respectively, and the second dilution retained (10^{-3}) two dishes contained 14 and 25 colonies respectively, so that

$$N = \frac{168 + 215 + 14 + 25}{[2 + (0.1 \times 2)] \times 10^{-2}} = \frac{422}{0.022} = 19182$$

$= 1.9 \times 10^4$ per ml or per g rounded to two significant figures.

Only two significant digits should be used in reporting the SPC. They are first and second digits (starting from left) of average of counts. Other digits should be replaced by zeros. For example, if actual count is 623,000 report the count as 620,000 ($62x10^4$). If third digit from left is 5 or greater add one unit of second digit (rounding off). For example, if actual count were 92,700, report the count as 93,000 ($93x10^3$). Significance of count can be assessed by determination of 95% confidence limits (that is, the range in which true count should be in 95 cases out of 100). This range is given by where $n\bar{x}$ is total number of colonies counted on all plates at chosen dilution and n is number of plates at that dilution ($\bar{x}$ is the arithmetic mean of colony counts at chosen dilution.

Spreader

Spreading colonies (e.g. Bacillus colonies spreading over agar surface) may cause trouble because of suppression of growth of other organisms as well as causing possible making of small colonies. Therefore plates containing spreading colonies should preferably not be used for counting. However, if there is no choice but to use them the presence of spreading colonies be noted in record. There may be error while preparing plates due to improper mixing of fluid and Agar, in such case reduce air, moisture inside incubator. Sometimes compact cluster of very small colonies occur at agar glass or agar plastic interface on bottom of dish, it is customary to count each cluster as one colony as it is usually assumed that they derive from individual organism contained within a clump (e.g. staphylococcus or Micrococcus) or chain (e.g. sterptococcus) which have spread at an early stage but which in depths of medium would have acted as a single viable unit and resulted is only one colony.

Replatability and Personal Error

A second count on a given plate should be

(I) Within 5% of the first when done by the same person
OR

(II) Within 10% of the first when done by another person or persons. If counts differ by more than these limits the cause is sometime failure to recognise minute colonies as such or more commonly, failure to differentiate colonies from food particles.

The Surface Spread Plate Method Counting TPC
Procedure

1. Add 15ml of melted, cooled (45° – 50°C) Plate count Agar to each petridish used and allow it to solidify. Dry agar plates preferably at 50°C for 1.5-2 hrs. If the plates are prepared in advance, the plates should not be kept more than 24 hours at room temperature or 7 days in a refrigerator at 2°–5°C.
2. Prepare food samples as described under the heading-Preparation and dilution of the sample of Food-page 23.
3. Using only one pipette, transfer 0.1ml of each of the dilutions to the agar surface of each of two plates. Test at least three dilutions even if the approximate range of numbers of organisms in food specimen is known. Start with highest dilution and proceed to the lowest, filling and emptying the pipette three times before transferring the 01.ml portion to the plate.
4. Promptly spread the 0.1ml portions on to the surface of the agar plates using glass spreaders (use a separate for each plate) a hockey shaped glass rod. Allow the surfaces of the plates to dry for 15 minutes.
5. Incubate inverted plates for 3 days at 29°–31°C.
6. Compute the surface spread plate count as described in the above method.

This method is not recommended for samples in which the numbers of organisms is likely to be lower than 3000 per g.

Horse-Blood Agar may also be used for the surface spread count.

The Drop Plate Method (Surface Colony Count)

This method described by Miles and Misra (1938) consists of placing drops (0.02ml) of serial dilutions on the surface of poured agar plates and counting the colonies that develop on incubation of the plates. It is not recommended for food specimens in which the number of organisms is likely to be lower than 3000 per gram. This method is useful when the bacteria are best grown in surface culture (e.g. when the presence of obligate aerobes is suspected) or when an opaque medium is employed. Surface counts are also indicated where the microbial population in a sample is likely to include bacteria that are killed by the brief exposure to 45° – 50°C that occurs when using melted agar for the pour-plate technique (Mossel, 1964)

Procedure

1. Prepare dried plates of plate count Agar or Horse Blood Agar as described in preparation of Agar plates. Prepare food samples as described in preparation and dilution of Food Homogenate in ringer solution as the diluent.
2. 0.02 ml drop of each dilution (10^1, 10^2.. 10^6) transfer with the help of Micropipette on six plates for each count. Allow the drop to be absorbed fully before inverting and incubating the plates. Complete absorption should occur within 15 minutes (longer times are indicative of inadequate drying)
   Incubate the plates inverted for 48hours at 29° – 31°C.
3. After incubation, counts all colonies for dilutions ranging from 20 to 100. The total count of six drops of the appropriate dilution is divided by six and multiplied by 50 to convert the count to colonies per 1 ml of the dilution used.

Use of Agar Droplets

Considerable savings in plates and media compared with the normal pour plate method can be obtained by performing colony counts on small droplets of Agar. Such a system has been described by Sharp and Kilsby in 1973 & 1971.

Procedure

A number of bottles or tubes containing 9ml of agar medium are held on the bench in a water bath at 45° – 50°C. 1ml of the samples or the first dilution of it is pipetted into the first bottle and mixed. A row of five 0.1ml drops of the inoculated agar is dispensed into a sterile Petridish, followed by 1ml into the next bottle of agar to provide the next decimal dilution. A standard Petridish can held three rows of five droplets and thus one dish allows the preparation of quintuplicate counts from three successive dilutions.

The principal disadvantages are that the system does not allow good differentiation of colony types when used media like brilliant green agar and smaller colonies at the edges of the droplets can be difficult to see.

USE OF DRY REHYDRABLE FILMS

Petrifilm consists of a sandwich of plastic films containing a disc of dehydrated medium. The film is placed on a flat and horizontal surface with its top layer peeled and an inoculum of 1ml of a dilution of sample is added to disc of dehydrated medium. This results in rehydration of medium. After replacing top plastic film, a plastic former is pressed down on to Petrifilm. This distributes the dilution across medium and produces a uniform depth of medium across disc. The Petrifilm is left on bench for few minutes so that it equilibrates.

Petrifilms are available coated with a number of media e.g. for counts of aerobic mesophilic bacteria (an AOAC official method, see AOAC 1955), yeasts and moulds, Enterobacteriaceae and Coliforms and Escherichia Coli. Main advantage of this system is the small amount of space taken by the films both in storage before use as well as during incubation.

Membrane Filtration

The membrane filter consists of a thin, very porous disc composed of cellulose acetate, cellulose nitrate or mixed cellulose esters. Membrane filters are available in a range of pore-size grades ranging from 10ηm to 8μm or more in pore diameter. Filters recommended for most microbiological isolations and counts are those with a pore size of 0.43-0.47μm. A few bacteria with a very small cell diameter require a filter with a pore size of 0.2-0.22μm.

Membranes can be obtained which carry a printed square grid to assist in the counting of bacterial colonies. These membranes allow large volumes of water or aqueous solution to pass through rapidly when under positive or negative pressure, but the small pore size prevents passage of any bacteria present. Bacteria which remain on surface of membrane can be cultivated by placing the membrane on an absorbent pad saturated with a liquid medium. Capillary pores in membrane draw up the nutrient liquid and supply each bacterium with nutrient. Bacteria will give rise to individual colonies after incubation. Media used are specially designed for membrane filter method and contain ingredients in concentrations different from those in standard media. Such media can be prepared in laboratory or obtained in dehydrated forms from e.g. Oxoid, Difco etc. Alternatively the membrane filter may be placed on surface of a set agar medium. In case of selective agar media their suitability for use with membrane filters should be determined, as different diffusion rates of medium constituents may significantly affect selectivity.

The membrane filter colony count has an advantage that provided the sample will pass through filter readily; small members of organisms can be detected in large amount of sample. Membrane filter counts thus may achieve sensitivity of multiple tube count whilst retaining accuracy of colony count method.

Alternatively, a microscopic examination can be carried out if bacteria are stained and the membrane filter then rendered transparently treating it with a liquid such as immersion oil or cottonseed oil. Suitable stains are loefflers methylene blue and crystal violet.

Membrane filters can be used for routine examination of water, air, sugar solutions, beverages, milk & milk products and spices. In milk & milk products first fat globules are broken down by use of an appropriate wetting agent such as iso-octylp-henoxy polyethoxy ethanol (Triton x-100)

Membrane filters are also used for sterilization of liquids including microbiological media and gases, the sterility testing of sterile liquids for clinical use, and the separation of phage from bacteria.

General procedure for performing colony counts is as follows:

1) Sterilization of equipment: Membrane filters should be wrapped in a paper and autoclaved for 15 minutes at 121°C. These should be kept in sterilized filter-holder. Between samples, funnel top can be sterilized by swabbing with alcohol and flaming.
   Dishes containing an absorbent pad should be wrapped and sterilized by autoclaving or in an oven.

2. Preparation of incubating dishes: To sterile absorbent pad in each incubating dish, add aseptically sufficient sterile medium to achieve saturation of pad without an excess of liquid being visible.
3. Filtration of a sample and incubation of the membrane: Attach sterile filter holder to a filter flask that is connected to a suction pump. Using sterile forceps, place a sterile membrane filter with its grid side up on the platform of filter unit, After first removing funnel top (take care not to contaminate interior of funnel). Replace funnel and lock into place.

Pour liquid sample into funnel and draw sample through membrane filter by applying suction. When sample has passed through filter, rinse funnel with sterile quarter- strength Ringer's solution. Slowdown suction and after removing funnel top aseptically, transfer membrane filter with sterile forceps to an incubating dish containing an absorbent pad previously saturated with an appropriate liquid medium. Filter should be placed on absorbent pad with a rolling action in order to avoid trapping air bubbles between filter and pad. After replacing lid of incubating dish, incubate with lid uppermost. Period of incubation normally required is some what less than that for poured plates.

After incubation, number of colonies is counted and viable count calculated per millilitre or gram of sample. Sometimes, there may be insufficient contrast between the colour of the colonies and the colour of the membrane, and in this case a staining procedure staining either the membrane or the colonies, may assist accurate counting.

Malachite green staining of the membrane: Malachite green of 0.010% is flooded incubated membrane for 3-10s, in this way, the membrane is stained. while the colonies remain unstained.

(b) Methylene blue staining of bacterial colonies: Saturate a fresh pad with a 0. 01% aqueous solution of methylene blue and transfer the incubated membrane bearing colonies to this pad for 5 minutes. Then transfer the membrane to a pad saturated with water. The methylene blue tends to be removed from the membrane more rapidly than from the colonies, and the colonies can be counted as soon as there is sufficient contrast with the membrane.

Most Probable Number (MPN) Counts

These counts are valuable as a general measurement of hygiene and the isolation and enumeration of coliform bacilli and Esch. coli in water and foodstuff.

Multiple tube count: It is common form of MPN Count. It provides an estimate of the number of living organisms in a sample that are capable of multiplying in a given liquid medium. The medium is selected which will support growth of bacteria under investigation. Sometimes, the count is based on production of gas and change of colour of medium used (e.g. coliform, Ecsh coliform counts determined by acid and gas production from Lactose present in MacConkey Media).

Procedure

1. Prepare food sample as stated under heading "preparation and dilution of the food Homogenate" *(page 23)*. Dilution blanks remaining from determination of plate count can be used (Caution: Do not delay to avoid growth or death of microorganism suspended in diluent).
2. A test tube of 10ml volume contains Durham fermentation tube in inverted state (75x10mm). The liquid medium (lauryl sulphate Broth) tubes of dilutions 1 in 10ml, 1 in100ml and 1 in 1000ml inoculate with 1ml of food homogenate.
3. Incubate tubes at 35° – 37°C for 24 hrs & 48 hrs.
4. After 24 hrs record tubes showing gas production. Return tubes not displaying gas production to the incubator for an additional 24 hrs.
5. After 48 hrs, record tubes showing gas production.
6. Select the highest dilution in which all three tubes are positive for gas production and the next two higher dilution e.g., if the last three positive dilution were 1:100, 1:1000, and 1:10,000 and the numbers of positive tubes in each dilution were 3,1 and 0, respectively, the results are recorded as 1:100=3, 1:1000=1 and 1:10,000=0. Report MPN=400.

If no dilution contains three positive tubes select the three highest dilution containing positive tubes. For example if the last four positive dilution were 1:10, 1:100. 1:1000 and 1: 10,000 and numbers of positive tubes in these dilutions were 2, 2, 1 and 1 respectively, the results are recorded as 1:100=2, 1:1000-1 and 1:10,000=1. Report MPN=200.

If further dilution beyond that showing three positive tubes were not prepared, select the last three dilutions made: For example, if the last three dilutions were 1:100, 1:1000 and 1:10,000 and the number of positive were 3,3 and 3 respectively, the results are recorded as" 1:100=3, 1: 1000=3, and 1: 10,000=3. Report MPN= >11000.

The most probable number (MPN) to result in this combination of positive and negative tube to be obtained by reference to probability table (see Appendix II)

It is to be noted that tables indicate MPN as the microbial concentration most likely to give that result, given the correctness of certain assumptions about for example the randomness of distribution of propagules in samples and dilution thereof. Some MPN tables indicate the 95% c.i. on the MPN Count, but once again this figure assumes that the experiment is in statistical control.

Generally five tubes are inoculated at each of a number of ten-fold dilutions. The MPN table for this arrangement is given in table appendix II. Cochran (1950) provided a sample technique for obtaining a rough estimate of the 95% c-1 as MPN/x to MPNXs where the value of X depends on both dilutions factor and number of tubes inoculated each dilution. For five tubes at each of a number of tenfold dilution x=3:30. For 3 tubes at each of a number of tenfold dilutions x=4.68. For two tubes at each of a number of tenfold dilution x=6.61. In fact, the 95% c.i. will depend in a complex manner on the numbers and positions of positive tubes. Accurate 95% and 99% c.i.s have been calculated by de Man (1975, 1977) and these are now incorporated into the MPN tables appended to ISO standards.

For determination of MPN, various broths used are Brilliant Green Lactose Bile Broth 2%, Laurel sulfate, Triptose broth and MancConkey broth. The multiple tube count method may be used in conjunction with selective and diagnostic media to determine the number of particular groups of organisms, for example, the number of coliform bacteria in food samples.

MPN Counts using the hydro phobic members filter (HGMF)

This method has been developed as a mean of obtaining estimates of counts of c.f.u.s (or propagules) with minimum dilutions. The ISO-Grid HGMF is a square membrane filter with filtration area of 5x3cm. This is divided into 1600 grid-cells by a black hydropholic print of grid-lines on membrane.

During incubation, microbial growth will occur within, and be bounded by, the grid lines surrounding a grid cell. Count can be made of number of grid cells after incubation or when a differential medium is used, the appropriate colour reaction. It is not a colony count, as a positive grid-cell can result from its being inoculated with more than one c.f.u. or growth unit. The HGMF instead provide an equivalent of MPN using replicate growth compartments at a single dilution. The MPN is calculated as $N \log \frac{N}{N - x}$, where N is number of grid cells (i.e. N = 1600 in case of ISO Grid) and x is the number of positive grid cells. The range of counts in sample inoculum capable of being estimated with reasonable precision by 1600 cell ISO Grid® is from 1 to about 5000. Thus fewer dilutions need to be prepared and tested them in the case with normal colony counting by either plates or membrane filters. Maximum precision occurs when half of the grid-cells are positive. This method has been recognised as an acceptable official method for the microbiological analysis of food (AOAC1995 & FDA1992).

Electronic counting devices are available for counting HGMFS.

An alternative to HGMF to achieve a single dilution MPN is to use a Microtitre plate to obtain multiple Inocula into wells of the plate.

In case a medium is designed such that after growth of the target organism results in a colour change, the microtitre plate can be examined automatically after incubation. Such systems are available commercially. One example is Quanti-Tray used in conjunction with colilert (IDEXX Laboratories, West brook, Main USA) to perform MPN counts of total coliforms and E.Coli.

The oxidation-reduction potential of a medium (Dye Reduction Methods)

These methods depend on the ability of microorganisms to alter the oxidation-reduction potential of a medium.

The activity of microorganisms is measured and not the count. The suitable indicator dyes used are methylene blue and resazurine. The time taken to reduce dye depends upon concentration and activity of bacteria present in sample, greater the number of bacteria, shorter is time taken to reduce the dye. However, many other factors of food such as nature, types of organisms present and medium used. The organisms must be capable of metabolism and growth in the medium to which dye is added, and if the sample itself is incapable of supporting growth, the dilution liquid should be nutrient liquid.

The dye reduction test is only applied to food or food constituents that have a reasonably constant and predictable composition.

These tests are based on determination of time taken for decolorization or the amount of reduction obtained after a set of time. This test indicates the freshness of the product. These methods are useful on the spot test for raw milks supplied in Diary. This test is less useful in assessing the quality of raw milk that has been refrigerated in bulk on the farm and then transported in a refrigerated bulk carrier. This test measures the ability of microorganisms present in the milk and its product like Ice Cream. Even in frozen state these microorganisms grow. The pathogens present do not affect the test result. It is difficult to assess the potential health hazard and the expected microflora.

ELECTROMETRIC METHODS

These methods are based on measuring the decrease in electrical impedance of growth medium as microorganisms metabolize and multiply. To measure impedance and calculating the growth, numbers of instruments are available manufactured by Malthus instruments, Don Whitely Scientific, Biomerieux and Sy-Lab. The software incorporated into the instrument monitor electrical changes automatically and interpret these changes into microbial load present in samples. A variety of microbiological criteria can be assessed by using appropriate selective media.

This system is useful when initial load is very high and measure it in shorter time.

The calibration curves for different microbiological criteria (e.g. aerobic mesophilic count, coliform count, Ecsh Coli Count etc.) can be judged in any type of sample. The instrument's computer is programmed to flag up on the visual display unit (VDU) the samples that had initial microbial counts in excess of maximum permissible limit under the relevant specification. In some cases within 6 hrs count can be measured. Thus it is useful in factories producing perishable and semi-perishable foods during the day. Any defective batch produced can be detected immediately.

Some microorganisms such as yeasts do not produce adequate changes in conductance in the culture medium. To overcome this difficulty, electrodes are not in culture medium but instead placed in an absorbent solution that detects the evolution of carbon dioxide from culture. The indirect conductimetric technique has also been used for detection of pathogen using selective media which are not susceptible to direct conductance measurements.

Determination of Total Number of Microorganisms in a food sample

The Breeds smear method for direct Microscopic counts: In this method the number of yeast, bacteria, for example milk, cell Suspension, broth culture can be determined by microscopic examination. The staining procedure does not differentiate between living and dead organisms, the bacterial count is known as "total count". The staining technique may be simple stain or Gram's staining method. The samples with high fat contents like milk, it is necessary to first defat (e.g. with Petroleum-ether). Newman's stain conveniently combines both defating and staining processes, although not allowing determination of the Gram staining reaction. The Breed's smear method is also used for counting animal cells-predominantly leukocytes which are found in much larger numbers in mastitis milk than in milk from healthy animals.

The microscope used first is to be calibrated so that the area of microscope field is known. A known volume (0.01ml) of sample or an appropriate dilution is then spread over a known area ($1cm^2$) on a glass slide. An alternative method is to draw two straight lines across the width of slide to enclose a length of 2cm. As a standard microscope slide is 2.5cm wide, this marks off an area of 5 cm^2, over which 0-05 ml of sample is spread. The sample is allowed to dry, then stained and examined microscopically. The average number of bacteria or cells per field is determined and it is then possible to calculate the number of bacteria or cells per millilitre of original sample.

MEASUREMENT OF THE AREA OF THE MICROSCOPIC FIELD

The unit of measurement used for measuring length and breadth of microbes is Micrometer (μm or popularly known as Micron) and Nanometer (nm also called Millimicron, mμ). A micron is one thousandth part of a milliliter. Therefore 1μm = 1/1000mm = 10^{-3}mm, 1nm = 1/1000μm = 10^{-6}μm.

1. The stage micrometer is adjusted with the help of x10 eye piece so that the graduated scale (1mm divided into 100 units of 10μm each) is in centre of field.
2. Place a drop of immersion oil on the stage micrometer and focus with the oil-immersion objective. Determine the diameter of microscopic field in micrometers (μm) using micrometer scales adjusting the tube length of microscope slightly if necessary. The same length is to be maintained in subsequent observations.

Calculate the area of the microscopic field in millimetres (mm) for the oil immersion lens and x 10 eye-piece using the formula: 77pier^2 = area of field where r is radius of field. Knowing the area of microscopic field, it is then possible to determine the microscopic factor (MF) by following calculation:

a) MF= number of fields in 1cm^2 (100mm^2) = 100/area of field in mm^2.
b) Average number n of organisms or cell per field = number counted/number of fields
c) MFxn= number of organisms or cells present in 1cm^2
d) Since 0-01 ml of sample was spread over 1 cm^2 or 0.05ml over 5 cm^2

e) MFxnx100 = number of cells or organisms present in 1ml. of sample

PREPARATION OF SMEAR

1. Mix the sample thoroughly and prepare dilution as necessary to produce smear with not more than 20 organisms per field. When pure cultures are to be examined, a barely detectable turbidity will be given by a suspension containing about 10 cells per ml.
2. With the help of micropipette transfer 0.01ml of sample to a clean glass slide and spread over an area of 1 cm2 marked by marker.
3. Dry the smear immediately by placing on a warm level surface or in an incubator at 53°C. Drying should be complete within 5 minutes to present possible bacterial multiplication.
4. Stain by appropriate method.
5. Examine slide by using oil-immersion objective.

Counting : Counting each single bacteria as one. Chains and clumps of bacteria are also counted as one. Bacteria further removed from the clump than the longest dimension of the constituent bacteria should be counted as separate. The direct count is also a clump count and bears a closer relationship to the probable count as each clump or chain would probably give rise to one colony only.

The accuracy of the count depends on the number of bacteria counted and as in the plate count technique, maximum practical precision would be obtained by counting about 600 items (bacteria) which is a fatigue work, it is recommended that a number of microscope fields are scanned and number of bacteria observed in each field summed, until the total number of bacteria observed is 1500 or more. Counting such a number will still provide acceptable precision. Note the number of field examined.

Determine the average number of organisms per field and calculate the number per milliliter or per gram of sample.

Direct Microscopic counts by Membrane filtration

The general procedure for filtration of sample has been explained under the topic Membrane filtration. After filtration remove the vacuum pump but do not remove membrane filter from filtration apparatus. Cover membrane filter with staining solution (e.g. Loeffer's methylene blue). Allow to stain for 5 min. and then start vacuum pump once more to remove staining solution. Pass distilled water, through membrane filter until the filtrate is colourless. Then remove the membrane and allow it to dry in air. When completely dry, saturate the filter with immersion oil by floating the filter on a small amount of immersion oil placed in a petridish. This procedure renders the membrane filter transparent. Remove the filter and mount on one or more microscope slides (in case of the usual 4.7 or 5cm filters it will be necessary first to cut the filter in half). Examine under microscopic examination using oil-immersion objective. Knowing the microscopic factor, the number of bacteria in a sample can be calculated from the effective area of membrane filter and the amount of sample passed through filter.

Direct Epifluorescent Filter Technique (DEFT)

Improved method of direct microscopic counting for certain types of samples is to combine membrane filtration, fluorescent staining and microscopy and known as epifluorescent filter technique (DEFT). A sample is filtered through a membrane filter which will retain all the bacteria, moulds and yeasts present. A fluorescent dye is then applied to stain the microbes and excess stain washed through the filter.

The membrane filter having specimen is connected with ultraviolet light from above. Microorganisms are seen flurorescing against a black background and is connected to image analyser to obtain counts automatically.

The precautions needing to be taken are: The use of suitable membrane (e.g. Nuclepore Poly carbonate membrane filters, amiable from Coring-Caster, High Wycombe, U.K.) prior filtration all staining and washing reagents through a filter particles, and use of non fluorescent immersion oil. DEFT Kits are available from 'Difco'. The method is generally used for milk samples, meat and poultry.

Flow Cytometry

A flow cytometer is an instrument that analyzes cells as they flow in a liquid stream through a beam of light. Flow cytometry techniques record characteristics for each cell flowing in the stream and are capable of analyzing thousands of cells in a matter of seconds. Characteristics may include size, shape, the ability to reflect or scatter light and fluorescence.

A liquid sample is passed through a flow cell illuminated by a laser and particles (in this case microorganisms) are detected automatically through a microscope focused on the flow cell. In food samples, it is necessary to detect and identify only those particles that are microorganisms. This is achieved by differential staining techniques. One such method is to use fluorescent dyes such as fluorescing isothiocyanate, fluorescin diacetate. Fluorogenic esters can be used to detect viable organisms, these are non fluorescent, but may be taken up by a viable organisms, and cleaned by non-specific esterases within the cell to release the fluorescent dye. For this method special instrument has been made by Chemunex (UK) known as Chem Flow apparatus for examining inter alia diary products including yoghurt, dairy deserts, fruit juices and soft drink, wine and beer. Now fluorescent labelled antibodies are available to detect specific microorganisms.

TURBIDIMETRIC METHODS

These methods are based on turbidity caused by microorganisms. Light is passed through suspension and absorbed light is measured. In addition the light-blocking power of an organism will depend on size, shape and transparency. The turbidity measurement can be coveted with microbial population only when a pure culture is examined and when the growth conditions of the culture have been standardized to ensure reproducibility of size and shape of organisms in population.

Applications of turbidemetric methods are for standardization of suspension of pure culture of bacteria to evaluate disinfectants or preservation of food staff. Such standardization ensures the bacterial concentration is within desired limits but total plate count will be necessary to know the actual count of the suspension. These methods are also applied for measurement of vitamins, other growth factors, antibiotic and other growth inhibitors, also for the determination of the effects of environment (e.g. temperature etc.) on growth.

The instruments used for measuring the amount of lights from a reference light beam as a result of passing through turbid is measured by Photoelectric calorimeters and Spectrophotometers, but these are insensitive to low bacterial concentrations. For low concentration growth Nephelometer is used to measure directly the amount of light scattered.

STANDARD OPACITY TUBES

This is a simple method for rapidly determination of bacterial concentration by comparing the turbidity (Opacity) of the suspension with the graded turbidities of a series of ten standard tubes. These tubes can be prepared by using barium chloride and sulphuric acid as under:

Tube no.	*10% Barium Chloride*	*10% Sulphuric acid*
1	0.1	9.9
2	0.2	9.8
3	0.3	9.7
4	0.4	9.6
5	0.5	9.5
6	0.6	9.4
7	0.7	9.3

The tubes used should be optically clean of standard diameter and standard wall thickness (nephelometer), add the above amounts of reagents. Stopper with glass standard joints, seal and keep refrigerated. Shake well before use.

METHOD OF USE

1. Transfer the bacterial suspension to a tube of dimension similar to the opacity tubes, to give a column 5-7cm high.
2. With the help of good light compare the sample tube with opacity tube one by one.
3. Select the opacity tube that most nearly matches the suspension.
4. The approximate number of organisms per milliliter is obtained by multiplying the number on the opacity by 300x106 in case of bacteria, or in the case of yeast by 10x106.

NEPHELOMETRY

The Nephelometers are used to measure turbidity of bacterial growth in a standard tube.

Procedure : The instrument is first warmed up as per instructions, then with help of distilled water zero is set. With the help of turbidity standard the instrument is set. The limit of sensitivity of nephelometers is about 10^5 organisms per milliliter. The sample under test is taken in borosilicate glass test tubes and reading taken. The same tube with standard turbid suspension is used and reading noticed. A graph is prepared by using a series of dilutions of known suspensions. From the graph test suspension tube reading will give the number of bacterial growth per milliliter.

Determination of buoyant densities of bacteria and food samples

The buoyant densities of food and shigella strains (any bacteria) after various treatments are determined by centrifugation in a self generated density gradient in standard isotonic percoll medium (SIM). SIM is diluted with peptone water (8.5g NaCl and 1.0g peptone 1-1Millipore Water to 60% of the original density of the SIM) and 6.5ml is added to a 13ml capacity, thick walled-polycarbonate tube (Beckman Instruments). Then 0.5ml sample, either the good homogenate or the bacteria in solution, is carefully layered on top. The sample tubes and one tube in which the sample volume is replaced by Density Marker Beads (Beads of known density, Pharmacia) in 0.15ml 10^{-1} NaCl are centrifuged on a centrifuge at 25000g for 25 min. The location of food and bacteria in the tube after centrifugation is recorded and the buoyant densities determined based on the calibration curve established from the location of the density markers. The food samples are homogenized for 1 minute in physiological saline (1 in 10 to 1 in 1 W/V) before density determination.

CULTURAL MEDIA, INGREDIENTS AND TESTING REAGENTS

Introduction: The Food material on which microbes grow is a culture medium and growth itself is a culture. Such a medium may resemble a natural substrate on which microorganisms usually grow (e.g. blood serum, milk, starch). Any medium most include all the necessary requirements for growth. It varies according to organism looking for. Main ingredients are (i) Water (ii) Nitrogen containing compounds like proteins, peptides, amino acids, nitrogen containing inorganic salts and sulfur compounds (iii) Energy sources like carbohydrate, peptides, amino acids, protein (iv) Suitable dyes are added to media to act as indicators to detect the formation of acid when fermentation reaction occurs or act as inhibitors of growth of certain bacteria but not to others (e.g. Phenol red which is red in alkaline or neutral medium but yellow in an acid one). Gentian violet inhibits growth of Gram +ve bacteria. The nutritional requirements of bacteria are different for different group of bacteria and it is not possible to formulate a single media for the growth of all types of bacteria. A medium like nutrient agar is capable to support the growth of many bacteria and also it can be used as a basal medium to which is added (e.g. blood serum or milk) to provide complex growth factors needed by more fastidious bacteria. B-Group vitamins are required by lactic acid bacteria which can be provided by addition of yeast extract.

A nutrient medium can be made selective or biochemically diagnostic by the addition of suitable compounds.

A few examples of non-selective culture preparation are as under:

NUTRIENT BROTH

This is a Media for general use to grow most bacteria (the most notable exceptions in food microbiology being lactic-acid bacteria and some fastidious pathogens)

INGREDIENTS

Peptone	10g
Meat extract	10g
Sodium chloride	05g
Distilled Water	1000ml

1. Weigh out ingredients and dissolve in water by just warming on hot plate.
2. Cool, adjust pH to 7.6 with 0.1N or N/10 Sodium Hydroxide (only 5 to 10ml is required).

3. Autoclave at 121°C for 15 min to precipitate phosphates.
4. Filter to get transparent medium.
5. Adjust pH with N Hydrochloric acid to pH 7.2.
6. Distribute in tubes, plug with cotton wool and cover the mouth with aluminium foil.
7. Autoclave at 121°C for 15 minutes and then cool to room temperature.

NUTRIENT AGAR

The ingredients are as used in Nutrient Broth, plus Agar: Nutrient Broth (pH 7.2 1000ml) and Agar 15g.

1. Dissolve 15g of agar in 1000ml Nutrient broth and autoclave at 121°C for 20 min.
2. Check that the pH is 7.2 and adjust if necessary.
3. Filter in a Buchner-type funnel by applying suction. Collect in a flask and distribute in test tube as required, plug with cotton wool and aluminium foil.
4. Autoclave at 121°C for 15 min or other relevant time. Check the pH.

DEHYDRATED MEDIA AND INGREDIENTS

The dehydrated media is dissolved in water; pH is checked and adjusted as per directions of manufacturers. Autoclave for 15 min at 15 lb pressure at 121°C cool at 45°C approximately and use for inoculation. Now the media in dehydrated form and also their constituents are available in dehydrated form.

INGREDIENTS FOR MEDIA

Most of the ingredients used to prepare media are listed below:

1. Agar-Prepared from Seaweeds and only bacteriological grade to be used.
2. BEEF Extract- Prepared from fresh cut pieces and dehydrated.
3. Bile Salts: Bile Salts is a standardized mixture of sodium glycocholate and Sodium taurocholate prepared from fresh ox-bile, for use as a selectively inhibitory agent in media.
4. Casitone: Casitone is a pancreatic digest of casein and is a rich source of amino acid Nitrogen. It is used in cultivation media

5. Dyes
6. Egg-yolk emulsion
   Eggs with intact shell and free from antibiotics are to be dipped in 10% aqueous solution of mercuric chloride, slightly warmer than eggs. Dry the shells with a sterile towel. Crack the eggs aseptically, roll the yolks on a sterile towel to remove the traces of white and then place the yolks into a stoppered sterile graduated cylinder. Add an equal amount of sterile saline solution 0.85% and mix well. Use the emulsion immediately. Discard the residual mercuric chloride in a manner to avoid contaminating sewage effluent and surface water. Ready made Egg-Yolk Emulsion is available from H1 media containing 3.5% Potassium Telluride.
7. **Gelatin:-** Gelatin is a refined water-soluble proteinaclous material free from fermentable Carbohydrates, used for solidification of culture media and for the detection and differentiation of certain proteolytic bacteria.
8. **Neopeptone:** Neopeptone is an enzyme protein digest suitable for use in the propagation of organism considered difficult to cultivate in Vitro.
9. **Ox-Gall :** Ox-Gall is a fresh bile purified for use in culture media as a selectively inhibitory agents in a number of bile media.
10. **Peptone:** This is a general purpose diluent and is recommended for the preparation of routine bacteriological media.
11. **Phytone :** Phytone is a vegetable peptone prepared by the papain digestion of soybean meal. It is used in media for the cultivation of fastidious organisms where rapid and profuse growth is required.
12. **Polypeptone:** Polypeptone is a mixture of peptone made up of equal parts of trypticase & thiotone(ingredient) for use in media where the characteristics of both peptones are desirable.
13. **Proteose Peptone:** This is specialized peptone prepared by the papain digestion of selected fresh meat for use in media mainly for the production of bacterial toxins.
14. **Skim Milk Powder:** Skim Milk Powder is prepared from raw milk after removing the fat and fat contents not more than one percent. It should be free from thermophile.
15. **Soytone :** Soytone is an enzymatic hydrostat of soybean meal that contains the naturally occurring carbohydrate of the soybean

and is suitable for media in which carbohydrates are not objectionable.

16. **Sugars:** All Sugars used in the preparation of culture media must be chemically pure and known to be suitable for bacteriological purpose.
17. **Thiotone :** Thiotone is prepared by pectin digestion of animal tissues and is characterized by a high sulphur content, making it useful in testing for hydrogen sulphide formation.
18. **Trypticase :** This is a peptone derived from casein by pancreatic digestion. A rich source of amino acid nitrogen and many uses in cultivation media.
19. **Tryptone :** Tryptone is a pancreatic or trypsinic hydrolysate of high-grade casein and a rich source of amino acid nitrogen.
20. **Tryptose :** Tryptose is a mixed peptone with nutritional properties making it suitable for use in media for the isolation and cultivation of fastidious organisms.
21. **Yeast Extract :** Yeast Extract is the water soluble extract of lysed yeast cells. It is an excellent source of growth-stimulatory substances.
22. **Water :** For the preparation of any culture media use only distilled water which has been tested and found to be non-toxic to microorganisms.

pH Determination and adjustment of cultural media.

The growth of microorganisms depends upon the pH of the growth medium. The medium required is acidic, alkaline or neutral and it is one of the selective factors for preparation of media. When the medium is made from its constituents it will be necessary to check and often to adjust the pH. Medium prepared from dehydrated medium often may not show the stated value. The medium after filtration and autoclaving must be adjusted to a predetermined value. For most unbuffered media autoclaving will lower the pH by 0.1-02 unit but occasionally the drop will be as great as 0.4. When buffering salts such as phosphates are present in media, the clear case in pH value as determined will be negotiable. Measurement of pH should preferably be made with a pH meter. If a meter is not available universal pH papers by Merck can be used.

Agar media must have their pH adjusted whilst molten and therefore at temperature above 45°C. Consequently the pH of the solidified medium should be confirmed at the temperature at which it will be used.

SEPARATION OF MIXED CULTURES

Streak plate method described at page 19 is applied to separate mixed culture into constituents organisms (provided the individual strains are present in approximately equal numbers). Isolated colonies are then picked up and subcultured as required.

Enrichment Method

The growing media is enriched to identify the target organisms by increasing its numbers. The physiological characters must be known about a particular organisms, e.g. optimum temperature or pH, nutritional requirements, tolerance of added inhibitors etc.

Sometimes pretreatment of the material itself is required such as heat treatment of suspension of the sample for the isolation of sporing or other heat-resistant organisms, followed by plating on the solid media. Optimum temperature of growth also helps in a mixed population and on incubation of material at the optimum temperature for the required organisms should increase its numbers in relation to other organisms percent. As the enrichment is done in liquid media, the particular enrichment procedure adopted is unlikely to yield a pure culture and the final isolation procedure still involves isolation of separate colonies on solid media.

Elective Media

A medium that satisfied the minimum nutritional requirements of the organisms concerned is known as an elective medium. It is useful for isolation of organisms in soil and spices and also for the organisms have unusual nutritional requirement e.g. wild yeasts can be isolated using lysine agar, on which organisms can not grow unless they can use lysine as a role source of nitrogen (Morris and Eddy 1957).

Selective Media

Selective Media is also an important method of separating mixed cultures. This medium may support growth of many types of organisms not required but due to presence of inhibitory agents, growth of organisms not required is restricted. Selective medium depends upon the nature of sample being examined. Inhibitory substances used in the preparations of selective media include dyes, antibiotics, bile salts and various inhibitors affecting the metabolism of enzyme systems of particular species.

Generally the selective medium is modified by adjusting the pH according to growth of only acid-tolerant or alkali-tolerant species. For examples yeasts, moulds and lactobacilli are acid-tolerant organisms and can grow on media at pH 4—5 where less acid tolerant organism are unable to grow.

Selective Media for Gram-negative organisms

Crystal violet used in a medium at a final concentration of 1:500000 inhibits the growth of many Gram positive bacteria (through not streptococci) while permitting that of Gram-negative bacteria (holding 1960). Penicillin incorporated at a final concentration of 5-50 units per ml inhibits many Gram-positive organisms and the medium thus becomes selective for Gram-negative bacteria.

Mac Conkey's agar is a selective medium used for the detection and isolation of coliforms. In this case the bile salt act as the selective agent, intestinal and coliform organisms being inhibited to a lesser extent than other organisms.

In some selective media a synthetic surfactant (Sodium dodecylsulphate) to achieve a similar selectivity is used.

Selective Media for Gram-positive organisms

Chemicals like sodium azide, potassium telluride, thallium acetate added to media to give a final concentration of 1:2000 to 1:10000 have been found to inhabit the growth of Gram-negative bacteria and these chemicals are frequently used in selective media for Gram-positive bacteria such as Glucose azide broth for example is used in the detection of faecal streptococci in water supplies and thallium acetate in a glucose agar is used for the isolation of lactococci from sour milk.

Differential Media

A differential medium is one in which certain species grow having characteristic colonies which can be easily recognized. On blood agar, haemolytic and non haemolytic species can be distinguished easily. In many cases, however, a medium may be both selective and differential. For example on MacConkey's Agar lactose fermenting Coliforms produce red colonies (as a result of acid production affecting the neutral red indicator) whereas non-lactose fermenting intestinal organisms such as Salmonella spp. produces colourless colonies.

IMPORTANT SELECTIVE MEDIA

MacConkey's Agar

This is of general use in the detection and isolation of members of the Enterobactiaceae It contains bile salt, lactose, neutral red, Peptone, Sodium chloride and Agar.

On this medium, after incubation for 24hrs, colonies of lactose-fermenting organisms are pink or red whereas those of non-lactose fermenting are colourless, typical appearance are as follow. E.Coli-red, non mucoid, Klebsiella aerogenes-pink, mucoid, Salmonella-colourless.

It is available in Dehydrated form. Modified MacConkey's agar is containing crystalviolet 0.001 g per litre and more refined effective bile salt at lower concentration. Violet red bile agar is a modification of MacConkey's agar containing 0.002 g per litre, addition of yeast extract, lowering the concentration of peptone and bile salt in more refined form.

Mac Conkey's Broth

It is a selective and differential medium used in the enrichment and detection of Coli and Ech.Coli (i.e. lactose fermenting Enterobacteriaceae) in, spices Meat, Milk and Water. It contains Peptone, Bilesalt, NaCl, Lactose, Bromocresol purple (10% Ethonolic solution) and Distilled water.

Presently a wide range of culture media are available commercially in the form of dehydrated media. These media are simply reconstituted by weighing the required quantities and b adding distilled water, as per the manufacturer's instructions. pH of the reconstituted media is checked conveniently either by pH papers or with the use of lovibond compratore with phenol red indicatore disc.

Method of adjustment of pH by Lovibond comparator

1. Take two clean test tubes and add 5ml of the medium to each of the tubes. One serves as a blank while phenol red indicator is added to the other tube.
2. Compare the colour of the medium with the phenol red indicator at the appropriate pH marking.
3. Add N/10 NaOH, drop by drop till the colour of the medium matches the colour of the disc at the required pH reading.
4. Calculate the volume of NaOH or HCl of 1/10 strength for 5ml of the medium to get the required pH.
5. Based on the calculation the volume of 1N NaOH or 1N HCl required for the total volume of medium can be calculated and added.
6. Check the pH of the medium once again before use.

AUTOCLAVING OF MEDIA AND DILUENTS

The constituents of media and especially Agar media is to be dissolved in distilled water and heat sterile in an autoclave at 120°C, 15lbs pressure for 15 to 20 minutes, this time begin when the "heating center" of the load reaches 121°C. A large volume of the medium or the diluent will also require longer time to reach 121°C say about 30min or more.

The container containing large volume of media is heated for longer time to reach 121°C, there will be thermal degradation as a result of extended heating time to reach sterilization. It is always advisable to use relatively small volumes (e.g. 100ml contents in 250ml flask) for sterilization.

METHOD OF SAMPLING FOR SELECTION, EXAMINATION AND ENUMERATION OF MICROBIAL COLONIES

Non-selective Media

Food having mixed populations of Micro organisms such as Meat, Milk, Soil, Spices and Water are plated on to non-selective media for estimation of total counts, enumeration of different kinds of Bacteria that develops on the plates. After growth it is possible to calculate the percentage distribution of various organisms present in the original sample.

Various methods have been devised to ensure that a truly random sample of colonies is obtained. In this exercise it should be appreciated that a minor but still possibly important component may be overlooked completely e.g. amongst population of 10^9 pseudomonads there may be 10^7 aeromonas but on average 100 colonies would have to be examined to reveal one aeromonas.

The particular method selected will depend on the numbers of colonies required for examination and the numbers of colonies present on the isolation plates. Some of the methods are as under:

a) Select every colony on a plate. It is a suitable method provided the numbers present are roughly equivalent to those required for examination.
b) Select every colony occurring either in a single sector or in opposite sectors of plate with the help of a feet-tipped pen divide the plate into quadrants. The plate may be supper imposed on a Harrison's disc and colonies are marked in each sector. It is a convenient method of obtaining a representative sample when only a few colonies can be studied.

Selective and Differential Media

Selective media will selectively permit the growth of target organisms (pathogens) and inhabit the commensals. Most selective media may differentiate the pathogens from other organisms capable of growth on the media.

For confirmatory tests colonies showing different characteristics typical of the target organisms may be chosen. The standard methods of the international organization of standardization (ISO methods) using selective media followed by confirmatory tests usually specify that five colonies be chosen from a plate. It is observed that this procedure of choosing only "typical" colonies for further confirmation and identification may lead to atypical strains being overlooked. For example sucrose fermenting and/or lactose, fermenting salmonellae are known to occur. Choosing only non-ferments from media containing lactose and/or sucrose will over look such strains of salmonella and their true significance and frequency of occurrence will be impossible to determine.

Anaerobic Culture Procedures

Anaerobic procedures usually measure not only obligate anaerobes, but also facultatively anaerobic organisms among the Enterobacteriaceae, Faecal streptococci and Staphylococci, unless selective media are used. To grow obligate anaerobes, cultures must be kept in an oxygen-free Environment. This is achieved by using a natural medium (e.g. Robertson's cooked meat medium) containing reducing substances (e.g. glucose, sodium thioglycollate, ascorbic acid, cysteine). Dissolved oxygen should be removed from a medium by heating. Excluding of atmosphere oxygen by sealing, by solidification of an agar medium in deep layers or by incubating the culture in an oxygen-free atmosphere. Now a days anaerobic jars, pre-reduced agar plates and deep agar in impermeable plastic bags are available. The anaerobic mesophilic count serves as an indicator of conditions favourable to multiplication of anaerobic food-poisoning organisms such as clostridium perfringens, especially in meat products, and it is necessary when examining food for this species in food poisoning investigations.

Cooked Food Medium

The Medium contains minced fat free ox heart in a Nutrient broth. The minced meat contains reducing substances which produce and help to maintain anaerobic conditions in the medium. This medium is helpful to grow anaerobes. Protaelytic activity of clostridium spororgenes is observed as it blackens and disintegrate the meat, also saccharolytic activity of Cl.perfringens by reddening of the meat particles.

Dehydrated form of this medium is available but it is seen that it is not reliable for cultivation of anaerobes. This medium prepared in the laboratory is satisfactory.

Preparation of Medium

Heat 1 litre of 0.05N aqueous sodium hydroxide to boiling and add 1000 g of minced fat free ox heart. Mix thoroughly, bring to boiling point, and allow to simmer for 20 minutes, stirring frequently. The reaction of the mixture should be about pH 7.5. Strain off the liquid and dry the meat by spreading on filter paper. Distribute dried meat in that tubes to a depth of 2.5-4 cm and add sufficient. Nutrient Broth or peptone water to give a total depth in the tubes of 4-5cm. Sterilize by autoclaving at 121°C for 20 minutes. After autoclaving tighten the caps/plug well. Immediately before use boil for 10min to remove dissolved oxygen and cool rapidly.

REDISSOLVED CULTURE

Tubes or bottles containing solid medium, are melted and used for culturing. Nutrient Agar with the addition of 1% glucose is the simplest medium.

1) Remelt the medium and maintain at 100°C for 10 minutes to expel any dissolved oxygen. Cool to 45°C and use immediately.
2) Add the inoculum and mix the contents of tube by rotation between the hands, air bubbles to be avoided.
3) Allow the Agar to solidify.
4) Incubate as per requirement,

The solid medium will hardly allow the oxygen to pass, the condition in the lower portion of the tube will be suitable for the growth of anaerobes. The method is also suitable for the growth of micro-aerophiles and facultatives anaerobes.

Semisolid Media

This media contains 0.2-0.3% Agar which helps to present the diffusion of oxygen into the medium. The important Media's are Tryptone Soyabroth, Nutrient broth with the addition of 0.5% Glucose (a reducing compound as well as being an energy source), 0.10% sodium thioglycollate and 0.02-0.2% agar to prevent convection currents, collected in deep layers in test tubes. 0.2 ml of a 1% aqueous solution of methylene blue per litre of medium is added as an indicator of oxidation reduction potential.

In nutrient broth, additional substances are added, mixed well and steamed until dissolved. The medium is then distributed in a suitable test tubes about 10-15ml capacity, plugged and sterilized by autoclaving at 121°C for 15minutes. The Methylene blue acts as an indicator of redox potential, being blue in the oxidized state and colourless in the reduced state. Should the medium show signs or turning greenish-blue after storage, it should be reheated before use to expel oxygen in water bath for 10 minutes.

Reinforced clostridium medium (RCM)

This is a semisolid medium which does not allow the growth of all obligate anaerobes, is useful for determination of anaerobes in food samples. It contains cysteine and glucose as reducing agents and when packed in 25ml screw capped. McCartney bottles, provides anaerobic conditions without the need for sealing. It is used for determination of viable counts of anaerobes. RCM Agar is available commercially in dehydrated form.

PARAFFIN WAX, VASELINE AND AGAR SEALS

Liquid media can be stored anaerobically by sealing the container with sterile sealing compounds such as paraffin wax and vaseline. The tubes containing the liquid media should be heated in boiling water for 10 minutes to drive off the dissolved oxygen and then sealed. The tubes are inoculated by means of a capillary pipette after melting the seal. The presence of clostridium perfringens in water or milk can be demonstrated in which the medium is sterile milk rendered anaerobic and surface sealed as above.

Anaerobic Jar

To create an oxygen free atmosphere for the cultivation of anaerobes, the plates or tubes containing any medium can be put in an anaerobic jar. Now a days different types anaerobic jars are available in market. One such type is made by Merck Which uses chemical to bind oxygen quickly and completely, creating an oxygen-free medium and a CO_2 atmosphere. Anaerobic jars to be used with external gas supplies have a lid carrying an inlet valve and an outer valve and preferably also have a built in manometer gauge. The air in the anaerobic jar may be replaced by oxygen-free nitrogen.

The alternative method is to make use of the catalytic non-explosive reaction between hydrogen and oxygen to remove free oxygen from the atmosphere inside the jar, leaving the cultures in an oxygen free atmosphere composed largely of hydrogen. Some manufacturers uses use cold palladium catalyst to bring about the reaction, the other uses a hot platinum catalyst. The precautions and uses are given by the manufacturers

REPLACEMENT BY NITROGEN

The cultures plates are placed in the jar and clamped, apply suction with a filter pump until the jar is evacuated to a pressure of 4cm Hg and then close the tap.

Fill the jar with oxygen free nitrogen to atmospheric pressure. Repeat evacuation and refilling twice more.

The level of residual oxygen after three cycles of evacuation and replacement by nitrogen will be approximately 0.003%.

SELECTION OF TYPE OF JAR TO BE USED

The microbiological requirement for example nitrogen filled jar is unsuitable for very oxygen sensitive anaerobes, but can be used with the more aero tolerant species.

CULTIVATION IN MICROAEROBIC AND CARBON DIOXIDE ENRICHED ATMOSPHERE.

Cultivation in Microaerobic atmospheres

Microaerophile campylobacter is grown in atmosphere containing 5-10% oxygen and about 10% carbon dioxide.

The required atmosphere can be obtained from gas cylinders containing gas mixture and filling the anaerobic jar. Gas Mixture generally used is 95% nitrogen and 5% carbon dioxide.

Cultivation in a carbon dioxide enriched atmosphere.

Some bacteria grow only or grow better in high concentration of carbon dioxide. Some anaerobes and micro aerophiles also prefer or require an enhanced concentration of carbon dioxide.

IMPORTANT BIOCHEMICAL TESTS FOR IDENTIFICATION OF MICROORGANISMS

This section deals with useful tests to characterize and identify the Bacteria.

These tests are based on the nutritional or osmotic requirements by organisms. In some cases a selective medium is used which allow the isolation of organism and also incorporate one or more biochemical tests

It is important to check that each batch of a medium is satisfactory by using a strain of bacterium known to give a positive reaction. The ninth edition of Bergry's manual of systematic Bacteriology lists the reactions of many type strains and sources of suitable culture. It is also useful to incubate uninoculated plate or tube with each test in order to detect false positive reactions due to impurities in, or a deterioration of medium or reagents.

Hydrolysis of Gelatin

a) Nutrient gelatin

Medium- Nutrient broth with the addition of 10-15% gelatin, final pH 7-2, Sterilize by autoclaving for 20 min at 115°C.

Stab inoculate a tube of nutrient gelatin and incubate at 20-25°C for up to 30days. If growth is poor at 25°C, tube may be incubated at the optimum temperature.

Observation: Liquefication of the test medium when an uninoculated tube has remained solid indicate that hydrolysis has occurred. Record the shape and extent of the liquefied portion of the gelatin. When tubes have been incubated above 25°C, the culture must be immersed in iced water for few minutes before being examined for hydrolysis.

b) Frazier's gelatin agar (modified)

Nutrient Agar plus 0.4% gelatin, final pH 7.2, Sterilize by autoclaving for 20 minutes at 115°C.

Inoculate the plate containing dried medium by streaking once across the surface, incubate at optimum growth temperature for 2-14 days.

Test reagent: 1% hydrochloric acid.

Observation: Flood plates with 8-10 ml of test regents. Unhydrolyzed gelation slowly forms a white opaque precipitate with the reagent. Hydrolyzed gelatin appears therefore as a clearzone. Record the width of the clear zone in mm from the edge of a colony to the limit of clearing.

Hydrolysis of Casein

10% Skim-milk with agar, sterilize by autoclaving for 20 min at 115 °C. Thick milk agar medium may be made by mixing 10ml of hot sterilized 2.5% water agar with 5ml of hot sterile skim milk (giving a 30% of milk agar) immediately before pouring the plates.

Test reagent: 1% hydrochloric acid.

Observation: Clear zones which are visible after incubation of the plates are presumptive evidence of casein hydrolysis, flood the plates with 1% hydrochloric acid which is a protein precipitant. Record width of clear zone in mm.

HYDROLYSIS OF COAGULATED SERUM

Medium: Loeffer's serum, consisting of three parts serum plus one part glucose nutrient broth. Glucose nutrient broth should be sterilize by autoclaving in autoclaving before mixing with sterile serum in sterile flask, Then the medium distributed aseptically into sterile small screw capped tubes. The medium is then coagulated by keeping the tubes in slopping position and heating slowly at 85 °C. The medium may be sterilized by heating in the oven at 85 °C for 20min on each of three successive days.

Incubate the serums slope by streaking with platinum wire and incubate at optimum temperature for 2-14 days.

Recording/Observation: Visual examination will indicate whether hydrolysis of the coagulated serum has occurred.

Production of Indole from tryptophane

Medium: Test tubes containing about 5ml of Tryptone sterilized by autoclaving for 15minutes at 121 °C. Tryptone is rich in tryptophane which is one of the amino acids. Inoculate as for a broth culture, preferably from young agar slope culture, and ncubate at the optimum growth temperature for 2-7 days. Incubate, for E.Coli detection less than 24 hrs.

Test Reagent: For detection of Indole several reagents and methods are available but Kovac's reagent being more sensitive due to the higher solubility of the dye complex in the pentanol layer.

Observation: Add 0.5.ml Kovac's Indole reagent, shake tube gently and then allow to stand. A deep red or pink colour may be due to the formation of spatol. Yellow colour if appears indicates Indole negative

Production of Ammonia From Peptone or Arginine

Peptone water tube (peptone 1%, sodium chloride 0.5%) and incubate with a sterile control tube at the optimum growth temperature for 2-7 days.

Test Reagent: Nessler's reagent

Observation: Add 1ml of culture to 1ml of Nessler's reagent in a clean tube. An Orange to Brown colour developed shows the presence of ammonia. The sterile control tube should be tested at the same time and should turn pale yellow or show no reaction.

Arginine Broth

This medium is used for differentiation of Streptococci. Inoculate a tube of argeinine broth and incubate with a sterile tube at the optimum growth temperature for 2-7 days.

***Test reagent and observation*:** As for method (1) (Peptone water)

Thornley's semi-solid Arginine medium. This medium contains phenol red as a pH indicator.

It should be dispensed in 5ml screw capped bottles to a depth of about 2cm. It is used mainly for the differentiation of Gram-negative rods. Inoculate the bottle of medium and then seal the surface with sterile liquid paraffin. Incubate with a sterile control at the optimum growth temperature for 2-7 days.

Observation: Arginine is hydrolysed with formation of ammonia, results in alkalinity of the medium and change in colour to red from salmon pink. Pseudomonas and Aeromonas produce ammonia from argininie in sealed medium, whereas members of

Enterobacteriaceae do not. The former can grow anaerobically, by being able to use *arginine* for the generation of ATP without needing oxygen.

Decarboxylation of lysine, ornithine or arginine

Medium-Decarboxylases of these amino acids, the medium is Moller's medium. It contains a small amount of glucose to initiate growth, higher concentration of selected amino acids, and a pH indicator, with anaerobic conditions provided by a sterile paraffin layer. After an initial lowering of pH by acid produced from the glucose, metabolism of the amino acid leads to a pH reversion.

Inoculate the organism, with a straight wire into a series of four tubes containing Moller's decarboxylase medium. The series consists of a control (no added amino acid) medium plus, lysine, medium plus ornithine, and medium plus arginine. Ensure that the wire penetrates beneath the layer of liquid paraffin. Incubate at 25°C and examine daily for up to 7days.

Observing the Result

A positive result is indicated by the colour of the medium changing to violet after an initial change to yellow. Negative and control reactions are yellow in colour when argentine containing medium show a positive reaction, if ammonia is liberated in the medium (tested by the use of Nesser's reagent) and the organisms does not possess a urease, the reaction is due to argentine dihydrolase.

Production of Hydrogen sulfide

Hydrogen sulfide is produced from the decomposition of organic sulphur compounds (e.g. cysteine and cystine) or from the reduction of sulphur compounds (e.g. Sulphite or thiosulphate)

a) Cycteine and Cystine

Peptone water or Nutrient broth with Cycteine or Cystine 0.01% is used. Lead acetate paper sterilized is enclosed at the mouth of the tube containing peptone water with cysteine or cystine and incubated at optimum growth temperature for 2-7 days, together with an uninoculated control.

Observation and Results: Production of hydrogen sulfide blackens the leadacetate paper. If no blackening has occurred by the end of incubation period, add 0.5ml of 2NHCl and replace the plug and lead acetate paper immediately. If any sulphide has been produced but has remained in solution, the addition of the acid will cause the liberation of hydrogen sulphide. Treat the uninoculated control tube similarly to check possible false positives.

b) Ferrous Chloride Gelatin

Nutrient broth with addition of 10 to 15% gelatin is added to 10% ferrous chloride in hot state. The final concentration of ferrous chloride is 0.05%. In hot state the mixture is dispensed in to sterile narrow tubes, quick cooling and sealing with sterile air tight (e.g. rubber stoppers). Inoculate by streaking. Incubate at 20-25°C for 7Day. Production of Hydrogen sulphide is shown by blackening of the medium.

c) Kligler's Iron Agar

This is a complete medium containing 0.03% citrate and sodium thiosulphate which is used for the differentiation of member of the interobacteriaceae. After sterilization, the medium is slanted with a deep butt (3cm butt, 4cm slant), inoculate the butt by stabbing and streak the slant. Incubate at optimum temperature for up to 7 days. Production of hydrogen sulphide is shown by the blackening of the medium. The medium also contains lactose and glucose for the differentiation of organisms on the basis of sugar fermentation.

Triple sugar iron agar (TSI agar) has been used instead of Kligler's Iron Agar, to detect hydrogen sulphide production. However with some hydrogen sulphide producing strain of citrobacter proteus, hydrogen sulphide is masked in TSI Agar by sucrose fermentation, so Kligler's Iron Agar is more reliable test medium.

Production of ammonia from urea

Christensen's Urea Agar: The basal medium peptone water is distributed in bottles or Test tubes, heat sterilized and cooled to 50°C, 20% Urea Solution previously sterilized by filtration is then added to give a final concentration of 2%. The medium is slanted, allowed to set, and is then ready for use. Inoculate as a slope culture, basal medium control tube containing no added urea should also be inoculated at the same time to check that Ammonia is produced from urea and not from peptone. Inoculate at optimum growth temperature for one to seven days.

Observation

Urease production and subsequent hydrolysis of urea result in production of ammonia. Aeromonas produce ammonia from arginine in sealed medium, whereas member of entrobacteriaceae do not. The former can grow anaerobically, by being able to use arginine for generation of ATP without needing oxygen.

Decaroxylation of Lysine, Ornithine or Arginine

Medium: Decarboxylases of these amino acids the medium used is Moller's, medium. It contain a small amount of glucose to initiate growth, a higher concentration of selected amino acids and pH indicator with anaerobic conditions provided by a sterile paraffin layer. After an initial lowering of pH by acid produced form glucose, metabolism of the amino acids leads to a pH reversion.

Inoculate the organism, with a straight wire into a series of 4 tubes containing Moller's decarboxylase medium. The series consists of a control (no added amino acid, medium plus, lysine, medium plus ornithine, and medium plus arginine). Ensure that the wire penetrates beneath the layer of liquid paraffin. Incubate at 25°C and examine daily for up to seven days.

Observing the result

A positive result is indicated by the colour of medium changing to violet after an initial change to yellow. Negative and control reaction are yellow in colour when arginine containing medium show a positive reaction, if an ammonia is liberated in the medium (tested by use of Nessler's reagent) and the organism doesn't posses a urease and the reaction is due to arginine dihydrolase.

Triple Sugar Iron AGAR (TSI)

TSI is a screening medium used to identify Gram-negative bacilli on ability to ferment carbohydrate, glucose, sucrose, lactose and to produce H_2S.

Principle: TSI agar contains protein, lactose, glucose and sucrose, a sulphur source (Thiosulphate), an H_2S indicator (Ferric Ammonium Sulphate), and a pH indicator (phenol red which is yellow in acid, red in alkaline conditions). Bacteria that ferment glucose produce a variety of acids, turning colour of medium from red to yellow. Large amount of acids are produced in butt of the tube (fermentation) than on the slant (respiration). Organisms growing on slant produce alkaline products from the oxidative decarboxylation of peptone. The alkaline products neutralize small amount of acids present in slant but are unable to neutralize large amount of acids in butt. Thus the appearance of alkaline (red) slant and an acid within (yellow) butt after 24 hours incubation indicates the organism is a glucose fermentore but is unable to ferment lactose or sucrose. Bacteria which ferment lactose and/or sucrose, in addition to glucose produce such large amount of acids that the oxidation determination of protein that may occur in slant does not yield enough alkaline products to cause a reversion of pH in that region. Thus these bacteria produce an acid slant and acid butt. Gas production carbon dioxide and hydrogen, is detected by presence of cracks or bubbles in medium. When gas pockets forms hydrogen sulphide, gas produced as result of reduction of thiosulphate. H_2S is a colourless gas and can be detected only in presence of indicator such as Ferric Ammonium Sulphate and Lead Acetate paper. H_2S combines with ferric ions to form Ferrous Sulphate which is insoluble black precipitate. Reduction of Thiosulphate will occur only in acid environment therefore the blackening is only seen in butt of tube which increases pH and a change in colour in medium from yellow to pink or red and form ammonium carbonate.

Reduction of Nitrate

$$(NH_2)_2\ CO + H_2O = CO_2 + H_2O + 2NH_3 = (NH_4)\ CO_3$$

Urease

The medium employed is nitrate peptone water. Peptone water contains 0.02-0.2% Potassium Nitrate. The medium is distributed in tubes, each containing Durham tube, and sterilized by autoclaving for 15min at 121°c.

Inoculate as for broth culture and incubate together with a sterile control tube at the optimum growth temperature for 2-7 days.

Test reagents: Griess: Ilosvay's reagents (modified) Recording result: Add 1ml each of the two reagents to the culture and to the control tube. The presence of nitrite is indicated by the development of red colour within a few minutes. The control tube should show little or no coloration. Nitrite test strips may be used..

A negative result should be confirmed by the addition of a small quantity of Zinc dust to the tube. This reduces to nitrite, any nitrate still present. Thus the red colour indicates that some nitrate remains. If the addition of Zinc dust does not result in the development of colour, no nitrate remains, the nitrate having been reduced by the culture beyond the nitrite stage. The presence of gas in the Durham tube indicates the formation of nitrogen gas.

Action on Litmus Milk

Inoculate litmus milk and incubate at the optimum growth temperature for up to 14days.

Observation and recording result:- (1) Examine tubes daily and record any changes in the medium. A number of different reactions and combinations of reactions may occur involving (1) Lactose (2) Casein, (3) other milk constituents, changes which occur are (a) acid production- Change of colour of litmus to pink (b) sufficient acid clot the milk (c) Reduction of litmus and loss of colour may occur and also Gas may be produced and appear as gas bubbles in the medium.

(2)(a) Coagulation of the milk may occur as a result of proteolytic enzyme activity affecting the casein, the litmus colour remaining blue. This is known as sweet clot.

2(b) Hydrolysis of the Casein as a result of proteolytic enzyme activity causes clearing and loss of opacity in the milk medium. Proteolysis may also result in an alkaline reaction due to ammonia production.

3. Utilization of citrate in the milk medium result in the production of an alkaline reaction, shown by the change to a deep colour in the litmus medium.

PHENYLALANINE DEAMINASE

Phenylalanine is an amino acid that upon deamination forms a ketoacid, phenylpyruvic acid of the Enterobacteriacea, only member of the Proteus, Morganella and providiencia genera possess the deaminase enzyme necessary for this conversion.

Principal: The phenylalanine test depends upon the detection of phenylpyruvic acid in the test medium after growth of the test organisms. The test is positive if a visible green colour develops upon the addition of a solution of 10% ferric chloride.

Incubate a phenylalanine agar slope and incubate at optimum growth temperature (37 or 30c) for 24h. Also incubate a known positive culture (e.g. a proteus) and a known negative culture (e.g. *Escherichia coli*.

Result: Add four or five drops of 10% ferric chloride solution to the phenylalanine agar slope culture, and gently rotate the tube. A positive reaction is shown by a green colour developing layer within 5min (the colour will fade quickly)

CARBOHYDRATE AND OTHER CARBON COMPOUNDS BIOCHEMICAL REACTIONS

Starch hydrolysis

This may be tested in solid or liquid forms although starch Agar is more convenient. It contains nutrient agar and 0.2% starch. The best results are obtained by preparing layer plates which are prepared by pouring 10ml. of nutrient agar into each plate, allowing to set, and then overlaying this with 5ml. of starch agar.

Inoculate a poured dried plate of the medium by streaking across the surface and incubate at the optimum growth temperature for 2-14days.

Observations: Sprinkle 0.5% Iodine solution on the surface of the plates. Unhydrolysed starch form a blue colour with the iodine. Area of hydrolysis therefore appear as clear zones and are the result of amylase activity. Record the width of any clear zone in mm from the edge of the colony to the limit of clearing.

Reddish: brown zones around the colony indicate partial hydrolysis of starch (to dextrin)

Acids formation from Sugar, glycosides and polyhydric alcohol's:

Take peptone water or other medium add 0.5-1.0% substrate. An indicator is also added in the medium to detect acid production

The indicator may be Andrade's indicator 1%, phenol red 0.01% or bromocresol purple 0.00 25%, Durham tubes are added to detect gas production. Sterilize by autoclave for 30min on three successive days. Substrate that may be excessively decomposed by heat sterilization should be sterilized as 10% solutions by filtration and added aseptically to tubes of previously heat sterilized peptone water (basal medium) to give the correct concentration of substrate.

Incubate at the optimum growth temperature for up to 7days:

OBSERVATION: The substrate decomposes, acid and gas is produced, indicators show the change in colour. Andrade's indicator to pink, phenol red to yellow, and bromocresol purple to yellow. Durham tubes show the production of gas. On continued incubation some organisms can cause pH reversion as a result of producing ammonia from peptone. Total utilization of glucose by bacteria that produces very little acid can be detected by the use of glucose-indicating test strips. The test strip changes colour if glucose is present.

DIFFERENCE BETWEEN OXIDATION AND FERMENTATION OF CARBOHYDRATES

Basal medium of Hugh and Leifson's is used to differentiate and identify Gram-negative bacteria on the basis of fermentative and oxidative metabolism of carbohydrates. Various workers have used this medium to classify unusual and certain non-fermentative Gram-negative bacteria. Bacteria produces acids from both oxidative and fermentative metabolism of carbohydrates. It is sometimes modified for particular group of organisms, such as lactic acid bacteria and staphylococci and micrococci.

The basal medium which contains a pH indicator (Bromthymal blue) is dispensed in 5-10 ml amount in 150x16 mm test tubes and sterilized by autoclaving. The carbohydrate is prepared separately from the basal medium as a 10% solution and sterilized by autoclaving or filtration. The sterile carbohydrate solution is added aseptically to the sterile melted basal medium to give a final concentration of 1%.

For each carbohydrate, stabe-inoculate two tubes of medium. Cover the surface of the medium in one tube with sterile liquid paraffin. Incubate at the optimum growth temperature for up to 14days.

Observation: The tube show change in colour from blue to red due to acid production. Fermentative organisms produced acid in both tubes. Oxidative organisms produce acid in open tube only and usually, or at least initially, only at the surface of the open tube

Methyl Red Test

Glucose phosphate broth is used as a medium after inoculation, incubate at the optimum growth temperature for 2-7 days.

Methyl Red Solution of 0.5% in Ethanol (0.1g Methyl Red Dissolved in 300ml of 95% Ethonal made up to 500ml with distilled water.

Observation: Add 5 drop of methyl red to 5ml of culture. A red colour at pH 4-5 or less is taken as positive. A yellow coloration is recorded as negative.

Voges Proskauer Test

This is a test for the production of acetyl methylcarbinol from glucose. To the inoculated medium after incubation, alkali is added in the presence of which may acetyl methyıcarbinol present becomes oxidized to diacetyl. The diacetyl will combine with arginine, creatine or creatinine to give a rose coloration.

Procedure: Inoculate glucose phosphate broth and incubate tubes at 35-37c for 48hrs. Pipette 1ml. of each culture to a separate empty culture tube and add 0.6ml. of 5% ethanolic solution of x-nepthol and 0.2ml of 16% solution of potassium hydroxide.

Shake the tubes, let them stand 2-4 hours and observe result.

Observation: The development of a pink to crimson colour at the surface, usually within 30 minutes, constitutes as a positive test.

SODIUM CITRATE TEST (UTILIZATION OF CITRATE AS THE SOLE SOURCE OF CARBON)

Procedure: Inoculate tubes of Koser citrate broth or simmon's, citrate agar: This contains bromthymol blue as a pH indicator, and agar. It is used as a slope, with a 1-inch butt. Use a straight needle and a light inoculum. Since transfer of nutrients with the inoculum can invalidate the test.

Incubate Koser citrate broth at 35-37°c for 72-96 hrs. Incubate Simmon's citrate agar at 35-37°c for 48 hrs.

Observation: For either medium the utilization of citrate and growth on citrate agar surface results in an alkaline reaction, the bromthymel blue indicator in the medium changes from green to bright blue and no visible growth as a negative reaction.

PRODUCTION OF POLYSACCHARIDE FROM SUCROSE

Sucrose agar containing nutrient agar and 5-10% of sucrose, sterilized by autoclaving.

Inoculate a poured dried plate by streaking to obtain separate colonies and similarly on a control plate containing 0.1% sucrose. Incubate control plate containing 0.1% sucrose, Incubate at 20-25°c or at 35-37°c for 1-14 days.

Observation: Synthesis of dextrin or leavan from sucrose is indicated by the development of growth of a mucoid character.

PRODUCTION OF CARBON DIOXIDE FROM GLUCOSE

Gibson's semisolid medium consists of 10% tomato-juice, 4% skim milk, 1% nutrient gar, 0.25% yeast extract and 5% glucose, final pH 6.5. The medium is distributed in tubes to give a depth of 5-6cm. Lactic acid bacteria requires Mangnese which is given by tomato juice, in place of that 0.4% $MnSO_4 4H_2O$ added to give about 1-10 ppm Mn++

Dissolve the media by heating at 100°c until molten, then cool to 45°c. Inoculate with young broth culture, cool in tap water, when set, pour into the tube molten nutrient agar at about 50°c to give a layer 2-3cm deep above the surface of medium. Incubate at 34-37°c for 14days.

Observation: The semisolid medium and agar seal trap any *carbondioxide* gas produced in the medium. Agar seal shows crackes. It may be necessary to place the tube in hot water to release gas.

REACTIONS INVOLVING LIPIDS, PHOSPHO LIPIDS AND RELATED SUBSTANCES

1. Hydrolysis of Tributryrin

The medium consists of yeast extract agar at pH 7.5 with the addition of tributyrin, emulsified and sterilized by steaming for 30 min. for three successive days. The medium is solidified in plates, 5ml. of molten tributryrin agar being overlaid on to previously poured and set plates of yeast extract agar. The medium is also commercially available in ready made form

Inoculate the poured plate of the medium by streaking on the surface, incubate at 35-37°c for 2-14days.

Observation: The clear zone appears on the surface resulting hydrolysis of tributyrin. Record the width of zone in mm from the edge of the colony to the limit of clearing. The reaction is usually regarded as being specific for lipase.

2. Hydrolysis of butter fat, olive oil and margarine.

The medium consists of yeast extract agar, pH 7.8 and butter fat or olive oil up to 5%, The medium is emulsified by shaking vigorously. The medium is poured in the plates. Inoculate a poured dried plate of the medium by streaking once across the surface and incubate at the optimum growth temperature for 2-14 days

Test Reagent: Saturated copper sulphate solution

Observation: Flood the plate with 5-10ml of reagent and allow to stand for 10-5 minutes. Drain off the reagent and wash the plates gently in running water for one hour to remove excess copper sulphate. A bluish green colour zone appears where lipolysis has occurred due to the formation of insoluble copper salts of the fatty acids set free on lypolysis.

3. Victoria blue butter fat agar and Victoria blue Margarine Agar

The Media contains a Victoria blue as an indicator of free fatty acids. Some strains of bacteria are inhibited by Victoria blue, so these media are not suitable for primary isolation of lipolytic organisms from mixed population.

Observation: When Lipolytic activity takes place, the free fatty acids combine with the Victoria blue to form blue salts. Deep blue zone surroundings or beneath microbial growth are thus an indication of lipolytic activity. The background colour of the medium should be pinkish-mauve.

HYDROLYSIS OF TWEEN COMPOUNDS

Polyoxyenthylene sorbitan monosters of fatty acids are Tweens, it also contains soluble calcium salt (usually calcium chloride). Any fatty acid released during incubation react with calcium salt and form a precipitate. Tween 20 (a laurie acid ester), Tween 40 (a palmate acid ester) Tween 60 (a stearic acid ester) and Tween 80 (an oleic acid ester) are the Tweens that can be used, but Tween 80 (an oleic acid ester) are the Tweens that can be used and usually chosen.

Inoculate a poured dried plate of the Tween Agar medium by streaking once across the center and incubate at 34c-37°c for 1-7dys.

Observation: Opaque zbnes surroundings the microbial growth are indicative of a positive lipolytic activity. Tween /Water mixture probably provide suitable conditions for the attraction of both lipases and esterases enzymes.

HYDROLYSIS OF LECITHIN

Egg.Yolk contains lecithin medium is prepared with a nutrient agar. Sodium chloride up to 0.9% and egg-yolk emulsion to 10%. Inoculate the medium plate dried after pouring by streaking once across the surface and incubate at 3 -37°c for 1-4days.

Observations

Hydrolysis of lecithin results in the formation of opaque zones (Lecithinase activity) around the region of microbial growth.

Egg-yolk containing sodium chloride 0.9% and 10% (v/v) egg-yolk emulsion. Inoculate and incubate at 34-37°c for 1-4days.

Observation: Lecithinase activity results in opacity in the egg-yolk broth medium, usually with thick curd.

Tests for the presence of active Enzymes

Catalase Test

Catalase is an enzyme used to break down hydrogen peroxide which is formed during aerobic respiration. It is toxic.

$$H_2O_2 \xrightarrow{\text{Catalase}} 2H_2O + O_2 \text{ (gas)}$$

The enzyme catalase is used to distinguish streptococcus (catalase negative) from staphylococcus and Micrococcus (Catalase positive) obligate anaerobes (e.g. clostridium) are usually catalase negative.

Test Reagent

Hydrogen peroxide (10,vol. concentration i.e. approximately 3%) Freshly prepared.

Method

(a) Pour 1ml. of hydrogen peroxide over the surface of an agar culture or a loopful of growth may be emulsified with a loopful of hydrogen peroxide on a slide.

(b) Place 1ml of hydrogen peroxide in a small clean test tube and add 1ml of culture taken aseptically from a broth culture.

Observation

Effervescence, caused by the liberation of free oxygen as gas bubbles indicate the presence of catalase in the culture under test.

OXIDASE TEST

The enzyme oxidase is used to differentiate pseudomonads from certain other Gram-negative rods, Most G+ organisms are oxidase negative.

Reagent: 1% aqueous solution of tetramethyl p-phenylenediamine hydrochloride as the reagents. The solution should be kept in- dark bottle in refrigerator but on long preservation it changes to purple and be discarded. Auto oxidation can be retarded by addition of ascorbic acid 0.1%.

The solution is irritant and toxic, use of commercially available test strips from Difco and oxide are available and can be used safely.

Method: Pour the reagent on the surface of agar growth in a petridish then smear bacteria on the same spot, if colour changes, the reaction is positive

RESULT- PINK-MAROON-BLK (Within 10- 30 minutes)

Coagulase Test: The enzyme Coagulase differentiates pathogenic (e.g. Staphylococcus aureus) from non pathogenic Staphylococci

Method: (a) Slide Method

Test Reagent: Dried rabbit/human plasma, reconstituted and diluted 1 in 5 has been found satisfactory. The test is carried out on 18-24 hrs nutrient agar cultures.

Make a slide into two sections with a grease pencil. Place few drops of normal sline 0.85% sodium chloride in a aqueous solution in each section. Emulsify a small amount of an 18-24 hrs agar culture in each drops placed on the slide until a homogeneous suspension is obtained. Add a drop of reconstituted dilute rabbit plasma to one of the suspension and stir for 5 seconds.

Observation: A Coagulase positive results is indicated by clumping which will not re-emulsify. The other section suspension remains as it is and serves as a control.

(b) Tube method: 18-24 hrs nutrient broth culture is used.

Reagent—as above

Method: Place 0.5ml of reconstituted diluted plasma into each of two small test tubes. To one tube add 0.5 ml. of an 18-24 hrs broth culture

Incubate both tubes at 37°c and examine after 1 hr and at interval for up to 24 hrs,

Observation:: Clotting indicates that strain under test is positive, it takes place within 1-4 hr. The other tube does not show clotting. The tube test is rather more reliable than the slide method.

OTHER IMPORTANT TEST

Phosphatase Test

Coagulase-positive staphylococci usually produce the enzyme phosphatase, can be detected by cultivation on a nutrient agar medium containing 0.01% phenolphthalein phosphate. It can be made more selective for staphylococci by adding polymyxin. This medium may be used to enumerate potentially pathogenic staphylococci in food stuffs etc. by using surface count technique or spread plates.

Medium: Liquefy the nutrient agar, cool to 50°c and add aseptically sterile 1% phenolphthalein phosphate solution to a final concentration of 0.01%, polymyxin .05%., mix and pour plates.

Reagent Concentrated ammonia solution.

Method: Inoculate the plates and incubate at 37°c for 24hrs. Expose plate to ammonia vapour by adding few drops of ammonia to a filter paper inserted on the lid of the plate.

Observation: Pink or red colonies indicate the presence of free phenolphthalein set free by phosphatase activity, shows phosphatase positive.

Hemolysis of Blood Agar

The hemolytic streptococci belong to some worlds more serous and devastating infectious diseases among both man and animals. Raw milk is the main source of streptococcol disease. The pathogenic streptococci are largely confined to the so-called beta-hemolytic species, it completely lyse the red blood cells and haemoglobin.

Medium: Nutrient agar containing 0.85% sodium chloride and 5% (v/v) defibrinated or oxalated blood. Horse blood is suitable for streptococci. Melt the nutrient agar medium, Cool to 50c and add the sterile blood (0.5ml. to 10ml. of agar) aseptically. Mix by gentle rotation and poured into the petridish.

Method: Streak the dried plate of the medium for staphylococci examination. If streptococci are being examined, either prepare pour plate or incubate, streak plate anaerobically. Incubate at 34-37°c for up to 2 days.

Observation: Clear zones around the colonies appear due to hemolytic activity. β-Haemolysis is the term given to this complete clearing of the blood agar when caused by streptococci. Zones of β-haemolysis possess sharply defined edges. Haemolysis is the term usually given to a greenish coloration produced around the colonies of some streptococci, it partially lyse RBC's and hemoglobin.

The terms haemolysin and β-hemolysin have a significance which depends on whether they are used with reference to streptococci or staphylococci. In case of staphylococci hemolysin produces clear zones and β-haemolysin produces dark hazy zones. When reading plates for type of hemolysis, be sure to look for colour changes in the surrounding agar. A common mistake is to look at the colour of the bacteria, rather than the agar.

Physiological Tests

Each organism has a minimum, optimum and maximum temperature, available water (aw), salt/sugar concentration, and pH for growth, survival and death of microorganisms in foods. The exact information on the growth, rate of growth depends upon the environment, These factors once determined, the shelf life of food can be accelerated, also spoilage of food can be prevented.

Growth rate Determinations on Pure Culture

Procedure: Take 100ml. of Culture medium in flasks and sterilize.

PHYSICOLOGICAL TEST

Temperature

Temperature is one of the most important environmental factors that regulates the growth of organisms. The survival of organism at a particular temperature is also important. The environmental temperature has also an effect on cell size, metabolic products such as pigments and toxins, nutritional requirement, enzymatics reactions and the chemical composition of the cells. The international commission on Microbiological specifications for foods (ICMF,1980a) divided into psychrophiles, mesophiles, thermophiles and psychrotrops.

Approximate temperature ranges of growth according to international commission on Microbiological specifications for foods is given below:

Growth Temperature (C)

Group	*Minimum*	*Optimum*	*Maximum*
Psychrophiles	-5 to +5	12-15	15-20
Mesophiles	-5-15	30-45	35-47
Thermophiles	40-45	55-75	60-90
Psychrotrops	-5 to +5	25-30	30-35

Organisms capable of growing at low temperature can be divided into two distinct classes. The first consists of those with a low optimum growth temperature are called psychrophiles (cold loving). It grows well at 0°c. Thy should produce a visible colony at that temperature in 7, 10 or 14 days.

The second class of organisms are those that have optimum growth temperatures 0°c to 5°e are called psychrotrophs. Most Microorganisms that grow in food at low temperature are psychrotrophic. The pseudomonads are an important group of psychrotrops. The spoilage of refrigerated food (meat, milk, eggs, fruit & vegetables) is generally due to psychrotrophic growth. Much of our supply held in cold storage which makes psychrotrophs important as potential spoilage organism. Mesophilies are important in food especially which causes food born illness, such as certain species of salmonella, staphylococcus, clostridum, Shigella and Bacillus. Although the rate of growth of microorganisms and enzyme activity is higher in mesophilic range, as compare to psychrophilic range even then the spoilage of food is faster in psychrophiles

The maximum temperature for growth of a microorganisms (i.e. the temperature above which multiplication ceases) is easier to determine than the minimum temperature to growth. The temperature at which multiplication ceases is often only a few degrees above the optimum, the growth rate falling sharply with a sharp cut-off point. However, at sub optional temperatures, the growth rate falls much more gradually.

THERMAL DESTRUCTION OF MICROORGANISMS

Many factors affect the heat resistance of Microorganisms. First different types of bacteria vary widely in their sensitivity to heat. Spore forming bacteria such as clostridium, botulism can survive several minutes even at autoclave or retort temperatures and pressure.. Vegetative bacteria are much less resistant to heat. Within the vegetative bacteria, Gram-positive bacteria such as Listeria monocytogenes are generally more heat resistant than Gram-negative's like salmonella. There can even be significant differences between species within the same genera and strains within the same species. Another factor that can heat resistance is the physiological state of the inoculum. The age of culture and condition under which it was grown can have a large effect. For example many studies have shown that sub- lethal heat shock can induce increased heat resistance. The matrix within which the organisms are heated can also generally influence heat resistance. Water activity is one of the most important factors. Heating a bacterial culture for increasing periods of time results in a progressive reduction in the viable count. The heavy load initially present takes longer time to kill the whole population at a particular temperature.

In the laboratory, we can take all of the above variables into consideration and conduct a thermal death-time study. A specific organism is chosen and packed into a specific matrix (broth, food emulsion, etc.). Several methods are used, but typically the inoculated matrix is sealed in a capillary tube or plastic pouch. It is heated to a specific temperature, held for a defined period time, rapidly cooled, then surviving cells enumerated. This sounds simple, but in reality these studies can be difficult to perform and should be done by competent and experienced man. Plotting the logarithm of the numbers of survivors against time gives a curve that tends to a straight line. The reciprocal of the slope of this line is the D

Value, which is defined as the time taken at a given temperature (T) to effect a reduction of 90% in the microbial population. The higher the temperature, the smaller will be the D Value. If the logarithms of the D Values obtained at various temperatures are plotted against temperature, and the best straight line is drawn through the points, the reciprocal of the slope of this line is the Z value. The Z value is defined as the number of degrees by which the temperature has to be raised or lowered to bring about 90% reduction or tenfold increase in the D value.

Using test tubes to determines D values and the Z values of an asporogenous culture:-

1. Dilute a'24-h broth culture (e.g. of Escherichia) to give a total count of about 3×10^8 per ml (This can be determined by microscopic count, or by nephelometry or opacity tubes if these have been calibrated using microscopic counts)
2. Put 10ml. of this dilution into 90 ml. of sterile diluent (e.g. 0.10% peptone water which contains number of glass beads. Shake well to mix and to break up lumps. Distribute in 5ml. amounts into five sterile 150 × 16mm test tubes)
3. Put the tubes prepared in (2) into a water bath at 58°c with a control test tube containing a thermometer and 5ml of diluent. When the temperature of the contents of the control tube reaches 1°c below the temperature of water bath, remove one of the tubes and start timing from this moment (To)
4. Rapidly cool the removed tube in cold water. Prepare six decimal dilutions and perform a plate count using all these dilutions.
5. Withdraw another tube and repeat step (4) after each of the following times 2 ½, 5, 7 ½ and 10 min.
6. Incubate the plates and determine the number of survivors after the various heating periods. Draw a graph of $\log_{10}$ (number of survivors) against time. From this determine the D58c (in minutes) as the reciprocal of the slope of the best straight tine.
7. Repeat steps (1) to (6) using a water-bath set at each of the following temperatures and removing tubes after-heating for each of the time shown
   56°C for 0,10,20,30,40 and 50 min
   54°C for 0,10,20,30,40 and 50min

52°C for 0,20,35,50,65 and 80min
51°C for 0,20,40,60,80 and 100min
50°C for 0,20,40,60,80 and 100min

8. Plot the graph $\log_{10}D_T$ against T, to estimate the Z value

Using sealed tubes to determine the D value and Z value of spores.

Using sealed tubes to determine the D values and Z values of Spores

The above method represents the simplest method for determining D and Z Values. It has many sources of inaccuracy and therefore the following method is described for use with spores of, for example, Basillicus stearothermophilus.

1. Glass tubes of 4mm external diameter, 2mm internal diameter, into 100-mm lengths. Seal one end of each tube with flame and plug with cotton wool and sterilize by autoclaving
2. Prepare a spores culture of B. Stearothermophilus by culturing on fortified nutrient agar in large sterile medical flat bottles, incubate at 55°c for 6days. The incubation temperature and period to provide a high ratio of spores t. vegetative cells should be determined by experiment for the strain under investigation.
3. Rinse the spore crop from the slopes by addition of a small amount of sterile phosphate buffer at pH 7-0, with a few glass beads to allow this to be effected using gentle agitation.
4. Add papain to the spore suspension to a concentration of 1mg per ml, and incubate at 37°c for 48hrs. to destroy vegetative cells.
5. Centrifuge the spore suspension, pour off the supernatant liquid and resuspend the pellet in phosphate buffer. Repeat this centrifuging and washing process six times, use sterile disposable tubes.
6. Finally resuspend the spores in 0.85% sodium chloride to a concentration of about 10^8 spores per ml (using counting chamber to determine the count microscopically) and store in a refrigerator.
7. Use 1ml of this working spore suspension to prepare a dilution series and carry out a plate count to determine the viable spore

population. Incubate the plates at 55°c with daily counting of colonies, avoiding contamination, until the count does not increase. This shows the viable spore population relative to the medium used (The plates incubated at 55°c or above should be placed in sealed bags or other containers to minimize dehydration of the medium)

8. Prepare a similar plate count after the suspended spores have been exposed to 80°c for 10min. A little heat treatment may be found to activate the spores and, by thus encouraging germination, to result in a higher count than that obtained on the unheated suspension.
9. This and following steps are carried out concurrently with step (8) that is, all the plate counts produced by step 8 to 14 are incubated together. Introduce the well-mixed spore suspension into eight of the tubes prepared in step 1, filling each tube to approximately one-quarter of its-length. The amount of spore suspension introduced into each tube can be predetermined by using a sterile microlitre syringe or by weight if uncaliberated. Pasteur pipettes are used. Immediately seal the open end of the tube carefully to protect the spore suspension from heating.
10. Use of one of the tube to ascertain the count at to. The surface of the tube should be Swabbed with 70% ethanol and also use sterile cutter. Aseptically place the tube in a sterile universal container with 10ml of sterile phosphate buffer. Break the tube using a strile metal rod. Mix the contents well and prepare a dilution series and note the plate count.
11. Immerse seven of the tubes in oil bath set at 110°C. Remove one tube after each of the following heating periods 1, 3, 5, 10, 15, 25 and 30 min.
12. Each tube should be cooled rapidly in a beaker of cold water, and a plate count made as described in step (10).
13. Incubate the plates at 55°c and determine the number of survivors per millilitre of suspension after the various heating periods. Draw a graph of log10 (number of survivors) against time, and ascertain the $D_{110°c}$.
14. Repeat the experiment using an oil bath set at each of the following temperatures and determine the number of survivors per millilitre of suspension after heating for each of the following periods.

115°C for 1,3,5,10,15,20 and 25 min.
118°C for 1,3,5,10,15,20 and 25 min
121°C for 1,3,5,7,10,13 and 16 min

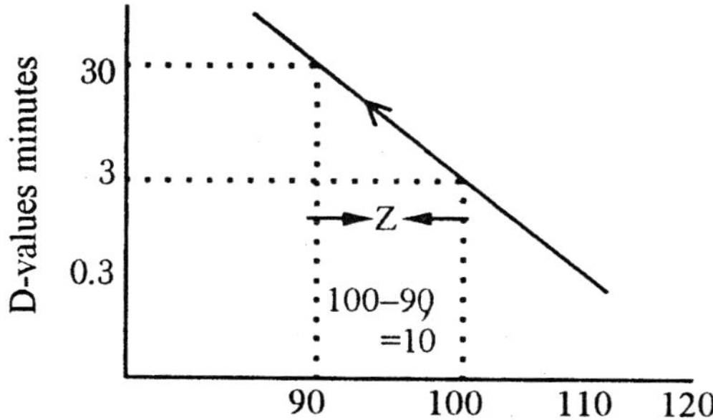

These times and temperatures have been found satisfactory in experiments involving B. Stearothermophilus.

15. Plot $\log_{10}D_T$ against T to estimate the Z value see the diag.

Use of D and Z Value in processing of Food products

Another terms that is used during process is F Value. In canning of food industry terms, this is equivalent time, in minutes, at 250°F of all heat considered with respect to its capacity to destroy spores or vegetative cells of a specific organism. The D and Z Value for an organism may be used to determine the destruction effect of a heat treatment on that organism.

A thermal death curve for this process is shown below, it is logarithmic process, meaning that in a given time interval and at a given temperature, the same percentage of the bacterial population will be destroyed regardless of the population present. For example, if the time required to destroy one log cycle or 90% is known and the desired thermal reduction has been decided (for example 12 log cycles than the time required can be calculated. If the number of microorganisms in the food increases, the heating time required to process the product will also be increased to bring the population down to an acceptable level. The heat process for pasteurization is usually based on a 12 D concept , or a 12 log cycle reduction in the number of this organisms. Thermal process for foods are designated as being five times D or 12 times and decided for each case

Note: Anywhere along the curve represent the same degree of thermal lethality

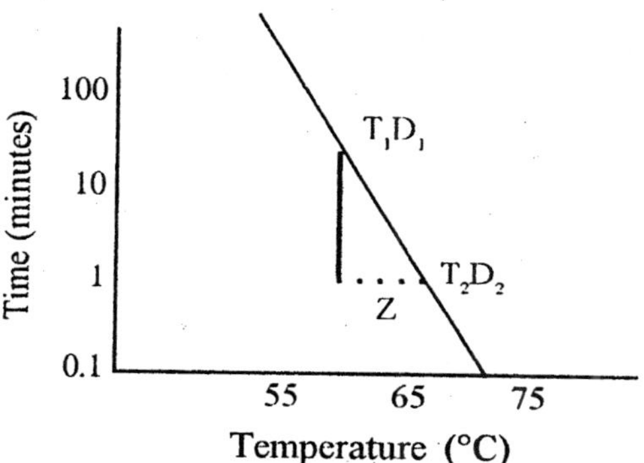

Thermal Death Time Curve for
Coxiella burnetti = $(T_2-T_1) = Z$ (40°c)

A working example of how to use D and Z values in pasteurization Calculations:-

Stored raw milk at the processing plant has bacterial population of 4×10^5/ml. Milk is to be processed at 79°c for 21 seconds. The average D value is 65°c for the mixed population is 7min. The Z value is 7c. How many organisms will be left after pasturization? What time would be required at 65c to accomplish the same degree of lethality.

Solution: At 79°c the D value has been reduced by two log cycles from that at 65°c since the Z value is 7°c. Hence it is now 0.07 min. The milk is processed for 21/60=0.35 min, so that would accomplish 5 log cycle reductions to 4 organisms/ml. At 65°c, you would need35 minutes to accomplish a 5D reduction.

pH TOLERANCE

The rate of growth of microorganisms is effected by pH of the medium. The buffering efficiency of liquid medium around its initial pH will have a profound significance on the organisms as they produce metabolites.

The pH tolerance of different organism can be observed qualitatively by using gradient plates. This method can established the approximate limiting pH for growth and this reduce the pH over which liquid media need to prepared for growth rate studies.

Method to study effects over pH 3.5-6.0

1. Plastic petridishes in square form about 25cm^2. Measure the working volume of the medium(10ml) required for these dishes. The amount of each of the two media to be added will be one half of this amount ½ 10ml.
2. Mark one end of each dish and draw lines 1cm apart across the back of the dish parallel to the marked end.
3. Place the dishes with marked end on a piece of rectangular rod of a suitable diameter so that when ½ 10ml of molten is poured into the dish it just reaches the marked end of the dish when a working depth of medium is achieved at the lower end of the dish.
4. Fill up each dish with ½ 10ml of an appropriate medium (e.g. trypton soyaagar)which has been adjusted to pH 6.0 with sterile 0.1 hydrochloric acid just before pouring into the plates. Allow to solidify.
5. Place the dishes on a horizontal surfaces and pour into each ½ 10ml of molten trypton soya agar adjusted to pH 3.5 with sterile acid prior to pouring. Allow to solidify. And then refrigerate for 18-24 h to permit equilibration between the two layers.
6. Take out the dishes, allow to reach room temperature.Set aside one dish for pH determination, streak other plates with cultures under investigation at right angle to the marked lines. Many cultures can be streaked on to each plate, provided the streaks are separated by at least 2cm apart.
7. Measure pH with pH meter fitted with flat tipped combination electrode, at this time measure the pH in each 1cm sector of the dish, taking measurements along the sector to determine whether there are any inconsistencies
8. Incubate the plates, after incubation determine the lowest pH at which each culture shows growth.

Importantt Note

The above procedure is described for the study of effects over a range of approximately pH 3.5-6.0 using mineral acids with no specific buffering being used. Organic filter sterilized acids can be used to adjust the pH, so that the responses of the micro-organisms to the organic acids can be assessed, for example if the lower pH be produced by adding filter sterilized lactic acid of 0.34% into the upper layer of Tryptone soya agar, this should give a range of pH across the plate about pH 4.6-7.0.

The phosphate buffered medium gradients plates having filter-sterilized 1.0M solutions of potassium dihydrogn phosphate, dipotassium hydrogen phosphate and ortho-phosphoric acid. Mix.buffer aseptically with molten tryptone soya agar at 1:10 just before pouring the plates. Using two of these three will provide different pH range e.g. with potassium dihydrogen phosphate in the upper layer and dipotassium hydrogen phosphate in the lower layer, the pH range will be from approximately 5-6 to 7.8 with potassium dihydrogen phosphate in the upper layer and orthophosphonic acid in the lower layer, the pH range will be from about 2 to 5.

SEROLOGICAL METHODS

The reactions between antigens and antibodies are the basis of Serological tets. Food Microbiological application of these reactions are of three types.

1. Classification and Identification of Microorganisms

In this case the unknown bacterium is reacted with known antibodies. This antigenic analysis is used in epidemological investigations and in systematic bacteriology. In the outbreak of food poisioning, Salmonella detection is important. The strain of Salmonella isolated from the food can be tested against a series of antisera containing antibodies with antigens of known serological types (serovars) of Salmonella. From faeces of patients Salmonella can be similarly tested and Salmonella in the food and faeces can be identified. Slide agglutination tests are used for this purpose (see under slide Agglutination test).

2. Detection and isolation of micro-organisms and their toxins

It is possibel to use immunological techniques (ELISA. Immunomagnetic & PCR) to detect target organisms in foods or in enrichment cultures, concentrate and isolate these organisms. The procedure is the speific antibodies will normally be linked or conjugated to an enzume lable or magnetic leads or Polymerase chain reaction (PCR).

The other procedure that can be used is affinity chromatoraphy in which the stationary phase such as agarose beads and one component such as antibody absorbs the reactant (in this case the spefic antigen) from the diluted food material (liquid phase). This procedure is used in a number of commercial kits for detection of specific mycotoxins (e.g. aflatoxins) in food samples. Detection and identification of antibodies in blood, milk and other body fluids.

A known bacterial cultures one used to detect the presence of specific antibodies against them. This type of tests one mostly used in clinical and veterinary microbiology. The concentration of antibodies depends upon an infection in the patient serum. A tube agglutination test is used to determine the concentration of antibodies. In this test a series of dilutions of series is tested against the microbial suspensions to determine the highest dilution of serum at which agglutination occurs.

SLIDE AGGLUTINATION TEST

In Food microbiology a test for detection of antibodies is the Brucella milk ring test used in the diagnosis of brucellsis in cattle. A suspension of killed Brucella cells stained with haematoxylin which provide known antigen is tested against the milk from the animal. If antibodies are present in the milk there will be a positive reaction (see the Milk Ring test for Brucella)

The agglutination takes place between cellular antigen and homologous antiserum. As a result of antigen-Antibody reaction, clumping of bacterial cells takes place resulting dense granules. In case of bacteria of the Enterobacteriaceae, the reaction between 'H'(i.e. flagellar) antigen and its homologous antiserum results in flocculent clumping whereas the reaction involving the 'O'(i.e. somatic) antigen results in a more dense and granular clumping.

Two types of antisera commercially available, polyvalent antisera which react with organisms of a particular serovars and other the use of which allows identification of a particular serovar

SEROLOGICAL METHODS

The slide agglutination test procedure

1. A glass slide (5x7-5cms) clear, grease free, mark it into the three sections about 1x2cm 'H', 'O' and C control) using a wax pencil.
2. Place one or two loops full of 0.85% saline solution in section 'C'.
3. Place a drop of a formaldehyde treated suspension of bacteria in the section marked 'H' and a Drop of a heat-treated bacterial suspension in section marked 'O'
4. Mix one loop full of undiluted polyvalent 'H' antiserum with the bacterial suspension labeled 'H'
5. Mix one loop full of undiluted polyvalent 'O' antiserum with the bacterial suspension labeled 'O'.
6. Tilt the mixture in three sections back and forth for 3minutes and observe against a dark back-ground if agglutination occurs there will be a clumping of the bacteria. Usually within 30 sec.

There should be no change in the control suspension.

LATEX AGGLUTINATION TEST

Latex beads(e.g. of polystyrene of dia 1μm) available commercially on to which antibodies have been absorbed. If the beads are white, the test is usually performed and observed against a black background. A more complex type of reagent uses a suspension of latex beads of more than one colour (e.g.red,blue and green) Each colour is associated with a different antibody or combination of antibodies. The agglutination is read against a white background. The colour of any flocculation indicates the presence of a particular serogroup . These coloured latex kits are available in the market.

Immunochromatographic kit one-use disposable plastic dipslide in which coloured particles are used in conjunction with chromatographic transport with positive antigen-antibody reaction appearing as a coloured line situated in an observation window where immobilized antibody is located. Such disposal dip slides usually used to detect the target organisms directly in enrichment cultures. Such kits are used to detect salmonella and Escherichia 0157:H7 and Listra manufactured by oxide.

THE MILK RING TEST FOR BRUCELLA ABORTUS AND BR MELITENSIS

Brucella abortus is the strain that most commonly infect cattle. B.suis and B, melitensis infections can occur if cattle are in contact with infected pigs or goats. Less severe disease occurs when cattle are infected with these stains. Adult cows are more susceptible than young stock.

The Milk ring test (MRT) demonstrates the presence of agglutinins (antibodies) in the milk, by the addition of stained bacteria to the milk sample followed by incubation for 30-60 min. The agglutinnins present in milk (cow's or ,sheep) will clamp the Brucella the stained agglutinin antigen complex rises with the fat globules This causes the cream layer at the top to become deeply coloured (in case of goat's milk the agglutinated stained bacteria go to the bottom of the test tube, owing to the different creaming properties of the milk).

The milk ring test is extremely sensitive and is very suitable for testing bulk of milk containing infected cow's or goat's milk.

METHOD: Add one drop of a stained Br.abortus or Br.melitensis suspension(available commercially) to a 25-mm high column of a well mixed milk sample in a small, narrow 75 × 10 mm test tube. Mix thoroughly by shaking but avoiding frothing. Incubate the mixture at 37°c for 30-60 min and then examine.

RESULT: The deeply coloured cream layer and the milk beneath the cream layer is white, the test is positive indicating the presence of agglutinins in the milk. If the cream layer is white and the milk beneath deeply coloured, or the cream layer is the same colour as the milk layer, the test is recorded as negative.

SEROLOGICAL TESTS

Enzyme-linked immunosorbent assays

The combination of enzyme linked immunosorbent assay (ELISA) and monoclonal antibodies (MeAb) have found widespraed use in food Microbiology. Elisa is most commonly preformed using MeAb coated microtitre trays to capture the target antigen. The captured antigen is then detected using a second antibody which may be conjugated to an enzyme. Addition of a substrate enables the presence of the target antigen to be visualized. Release of chromophore or fluorophore by the enzyme activity can be measured. The intensity of colour is proportional to the amount of antigen present. A wide range of enzyme-linked immunosorbent assays(ELISA) are commercially available especially for Salmonella, pathogenic E.Coil and Listeria. Immunoassays may be performed either as the standard ELISA tray (96 well) format, or the antibodies can be immobilized on to a solid support in form of a dip stick.

Concentration of Microorganisms by Immunocapture

The most successful applications at present are in the detection of target microorganisms in pre-enrichment or selective enrichment-broths. perhaps after a shorter than normal incubation period. To capture the target organisms uses magnetic beads coated with antibodies specific to the target micro organisms. The magnetic beads be stirred into an enrichment broth culture. After allowing time to permit binding of the target organisms to the beads, the beads are attached to a magnet placed against the side of the container whilst the liquid is removed. The beads are washed and can then be transferred to a solid selective medium to detect the presence of any organisms on the surface of the beads. Alternatively the bacteria-carrying beads can be examined by an impedance/conductance instrument. Impedance microbiology detects microbes either directly due to he production of ions from metabolic end products or indirectly from carbon dioxide liberation. The direct method monitors impedance (reciprocal of conductance) changes the growth of the medium. Microbial metabolism produces ionic end products (organic acids and ammonium ions) from the growth medium and therefore increases the conductivity of the medium.

The Precipitin Test

In this test a clear solution of protein or polysaccharide antigen extracted from bacterial cells is used. In Agglutination test, the antibody causes clumping of an already present bacterial cells. In precipitin test, the clear antigen solution when mixed with appropriate clear antiserum, will produce a mixture that first terms cloudy and then precipitates.

Lancefields's Test for Streptococci and preparation of antigen solution.

1. In yeast glucose lemon broth strain of streptococcus are grown:
2. Certrifuge the broth and discard the supernatant liquid.
3. Add 2ml of N/50 hydrochloric acid containing 0.85% w/v sodium chloride to the sediment of Bacteria and transfer the suspension to a small test tube, using a pipette.
4. Place the tube in boiling water and leave for 10minutes.
5. Cool the tube quickly under running water.
6. Add one drop of 0.04% phenol red solution the liquid should turn the phenol red to yellow.
7. Add drop by drop 0.3 N Sodium hydroxide until the fluid is alkaline (red in colour) Then add, drop by drop, 0.1 N hydrochlorie acid until the liquid is neutral (Salmon-Pink)
8. Centrifuge and remove the supernatant liquid by means of a pasteur pipette to another test tube, and discard the sediment. The liquid constitutes the antigen solution,which is ready for use. To preserve, add 0.5% phenol and keep in fridge.

The Precipitin ring test: This test is performed common only for streptococcal grouping. The quantitative estimation is done by using tenfold dilutions of antigen ranging from 1:10 to 1.10^7

Procedure

1. 0.01 ml of antiserum is placed in small test tube.
2. 0.01 ml of prepared antigen or of a serial dilutions of antigens is layered over the antiserum. This must be done carefully so that a sharp interface forms between the two solutions. A pasteur pipette with a finely drawn end is used and the tip is placed against inside of the tube just above the surface of the antiserum so that the antigen solution runs on to the antiserum.
3. Similarly prepare saline-antiserum control and an antigen normal serum control.

Results: Formation of a white precipitate at the interface of the reagent within 30 min. indicates a positive reaction. Standard antisera are available commercially.

3

Yeasts and Moulds

Introduction

Yeast and mould are widely distributed in nature and may be found as part of the normal flora of a food products, inadequately sanitized equipment or as air born contaminants. Yeasts and Moulds grow more slowly than bacteria in non-acid, moist foods and therefore seldom cause problem in such foods. However, in acid foods and foods of low water activity, they outgrow bacteria and thus cause large spoilage of fresh foods, cereal products, salted and pickled foods, frozen and dehydrated foods which are improperly stored. More so they produce mycotoxins, alter the substance, allowing for the outgrowth of pathogenic bacteria.

Genral Conditions for the Growth of Yeasts/Moulds

The media's used for yeast and mould differ from bacteriological growth media as requirement for growth is different. Most of fungi have an optimum pH much lower than that of most bactria, also. They are more capable of growing on media of inorganic salts with the addition of carbohydrate as an energy source. Some fungi require B-group vitamins, other nutrions which may need to be fulfilled by adding 0.1% yeast extract. Almost all moulds, particularly the saprophytes and are obligate aerobes

Media for the growth of yeasts and moulds

The classical method for the enumeration of yeasts and molds uses an acidified medium to inhibit bacteria. The important media used are as under:

1. Meat extract agar for isolation, counting and cultivation of moulds and yeasts. The pH may be at 5-4 or 3-5 depending on the purpose of the medium.
2. Potatoes Dextrose agar acidified with tartaric acid to pH 3.5 following sterilization, both for yeasts and moulds. It is necessary to incubate the plates at a temperature range 20 to 25°C for 5 days
3. Davis's yeast salt agar for counting moulds and yeasts. The pH of the media is 3.5 to inhibit bacterial growth.
4. Dich loran rose bengal chloramphenicol agar for the isolation of moulds from the samples containing large numbers of bacteria.
5. DG18 agar (dichloran 18% glycerol agar) (Pitt and Hocking 1985; Pitt et al. 1992) is suitable for the culture of osmophilic (xerophilic) fungi. However, malt yeast 50% slacoseagar has been recommended for the culture of extreme xerophilies.
6. Osmphilic agar (Scarr, 1959, Beech and Darexport 1969) is a wart agar with a high concentration of sucrose and glucose, for the growth of osmophilic and osmotic learnt organisms.

As an alternative to acidification media such as Malt extract agar can be rendered selective against bacteria by the addition of antibiotic chloramphenicol. Acidified media present certain shortcomings such as spreading mold colonies, occasional growth of bacteria, precipitation of food particles and inability of some yeasts and molds to grow at the low pH of the medium.

MOULDS EXAMINATION

From the incubated plate, examine the colonies under the X10 (low-power) objective of the microscope. Record the characteristic of the colonies.

Prepare slides of the mould growth in the following way. Pick off a portion of the colony with a needle and tease it out in a drop of lactophenol cotton blue placed on a microscope slide. Cover this position of the slide with a clear coverslip, taking care to exclude air bubbles.

Examine the prepared slide under the microscope using X10 low power and then using x40 (high power dry) objective for a closer examination of a selected field.

Moulds are examined also in wet preparation (e.g. placing a drop of water and mixing the colony in it) example of which are Fusarium, Mucor, Rizopus. Many moulds are inadequately wetted by water and become enclosed in air bubbles, consequently a mountant such as lactophenol is employed; Usually containing a stain such as picric acid or cotton blue. Lactophenol prepared slides can be preserved for longer times, although a gradual deterioration will occur from other causes.

Yeast Examination

Place a drop of water on clean slide, suspend a portion of culture in it. Add a small drop of Gram's Iodine and cover with a cover slip. Observe with the x40(high power dry) objective. Loeffer's methylene blue can be used instead of Iodine.

Produce a heat fixed smear in the usual way and stain by Gram's method.

Identification of Moulds and Yeasts

The unicellular form gives yeasts an advantage over the mycelial form of moulds. Yeasts may reproduce by budding, binary fission ascospores or, less commonly by other methods. Identification is more difficult than in the case of multicellular moulds, partly because yeasts rarely produce ascospores on ordinary media. Sporogenous yeasts can be induced to form ascospores by subculturing twice on a nutrient agar containing 5% glucose and 0.5% tartarie acid, followed by subculture on to an agar medium containing 0.04% glucose and 0.14% anhydrous sodium acetate only.

Descriptions about Yeasts and Moulds

Wet-mounted stained slides prepared from petri-dish cultures invariably causes disintegration of sporing structures and disintegration of pseudomycelia or maycelia that are fragmenting into arthrospores. It is therefore more satisfactory to examine yeasts and moulds by slide culture, when chances of fragmentations are very less. The slide culture methods for yeast is different than mould. Certain fungi such as Geotrichum, candida and Endomyces which are intermediate in their structure between yeasts and mycelia fungi, may at the initial stage show the structures either yeasts or moulds, depending on the cultural condition such as medium, temperature and period of incubation.

The number of media are to be tested in petri-dish culture for the growth of fungi. At least two media, one with high and one with low concentration of carbohydrate should be used. Cultures should be incubated at 25°c or other appropriate incubation temperature for 5 days and are to be checked every other day. These are examined using a stereoscopic Microscope with incident light. The spores of a number of moulds encountered in food may, if inhaled. either induce an allergic response or establish a lung infection, so carefully culturing of spores should be done.

Identification of Yeasts

Yeasts are identified on the basis of cultural and Morphological characteristics and production of ascospore and ballistospore. Some biochemical tests are important also. The generic level of yeasts isolated from food and dairy products, these tests help in identification, also yeast isolates yeasts grow on agar media (Malt extract Agar, Cornmeal agar), young yeast colonies grow in 2 to 5 days at 25°c and appear like dome with entire edges, butyrous consistency, malt smooth surface, white in colour but some time creamish, pinkish or solmon. On prolong incubation the colonies of some species tend to become wrinkled and of a dry and crumbly consistency. These colonies resemble those of some bacteria but due to low pH or selective media, hardly chance of growing the bacteria. Subsurface colonies of yeasts in pure plates often have a characteristic stellate appearance.

Microscopic examination of wet-mounted preparations, either unstained or stained with Gram's Iodine or Loeffler's spherical, ellipsoidal, avoid or binary fission. Slide culture method is the best for production of mycelium and the growth of yeasts should be done on Malt extract agar or potato extract agar.

Slide Culture Preparation for Yeasts

A petri-dish containing filter paper in the base, on which is a U-shaped glass rod carrying two clean microscope slides-all to be sterilize

Identification of Yeasts

After sterilization, aseptically, add a few millilitres of a sterile 20% solution of glycerol to the filter paper. The glycerol prevents from drying out slide culture during incubation. With the help of a pipette and molten Malt extract agar or Potato dextrose agar to the surface of the slides to form a thin layer of the agar medium. Inoculate the slides with the culture to be examined for yeast with streaking wire and place a cover slip over part of the inoculated medium. Incubate at 25°c for up to 7 days with the lid of the Petri-dish in place The slides are removed from petridish and examined the growth under the cover slip.

Physiological and Morphological test for Yeasts

Ascospore production: From slide culture as describe above, subculture on to gypsum block or V-8 agar.

Ballistrospore Production: Sterilize petri-dish containing a very thin U-shaped glass rod, filter paper disc in the base, and microscope slides. Cover only one slide with agar medium (malt extract or Potato dextrose agar). Inoculate the slide with yeast medium and invert the inoculated slide above the second slide with the thin U-shaped rod between the two slides. Replace the lid of the petri-dish and incubate at 25°c. After incubation examine for ballistospores presence or absence of the lower slide. Examine also the upper slide to determine the morphological and cultural characteristic of the yeasts.

Carbohydrate Fermentation

The tub containing Carbohydrate broth with inverted Durham tubes, inoculate with yeast. The carbohydrate fermentation broths should consist of 0.5% yeast extract with the addition of 2% substrate. A control containing no added carbohydrate should be used to check the efficiency of yeast extract. Incubate at 25°c and examine every day for 10 days.

Use of Ethanol as a Sole Carbon Source

The tube containing ethanol, inoculate very lightly with yeast under test and also a control without ethanol. Incubate at 25°c for 3 weeks, examine at frequent intervals for growth. Yeast extract solution is used as a source of Vitamins and growth factors, the growth of the yeast being examined must be compared in the tubes with and without ethanol for positive results.

IDENTIFICATION OF YEAST

Cycloheximide Resistance

Inoculate the yeast into filter sterilized yeast Nitrogen Base 0392 (Difco)containing 0.5% glucose and 100-g cycloheximide per ml. Incubate at 25°c for 4weeks. Examining weekly for growth. Good growth within 1 week is taken ad indicating high resistance. Scant or no growth in 3 or 4 weeks is takn as indicating high sensitivity.

Isolation Techniques for Mixed Cultures

All initially isolated yeast's may be contaminated or in mixed culture. Direct mounts and subsequent streaking for colony isolation are necessary to confirm the purity of each yeast isolate. Pure cultures are essential for assimalation procedures. Streaking subcultures for spatial isolation is adequate in most cases. The following additional techniques can be used to purify yeasts.

A. *Bacterial contamination*

1. Colony isolation on SAB agar.
   a. Suspend a small portion of the yeast to be decontaminated in sterile distilled water
   b. Streak a loopful of the suspension for colony isolation onto a plate of SAB agar.
   c Incubate for 30°c for 48 hrs.
   d Examine for isolated colonies. Verify purity with a direct exam
   e If colonies are not pure, further steps are necessary (step 2 follow)
2. Colony isolated on SAB Agar plus chloramphenicol, SAB plus pencillin and streptomycin & BH plus 10% blood plus gentamicin and chloramphenicol.
   a. Suspend a small portion of the yeast to be decontaminated in sterile distilled water.
   b. Streak a loopful of the suspension for colony isolation on to one of the above media.

c. Incubate at 30°c for 48hrs. Check for purity. If colonies are not pure, further steps are necessary (step 3 follow).

3. Acidification of SAB broth.

a. Suspend a small portion of the yeast to be decontaminated in sterile water.

b. To each of 4 tubes containing 10ml. of SAB broth, add 1 drop of 1N HCL to the first tube, 2 drops to the second tube, 3 drops to the third tube, and 4 drops in the fourth tube

c. Add 0.5ml. of the contaminated yeast suspension to each tube.

d. Incubate at 30°c for 24 hours.

e. Subculture a loopful of each broth to SAB agar plates. Sreak for isolation.

f. Incubate the SAB plates at 30°c for 48 hours.

g. Check purity. If still not pure, consult the supervisor

4. Mixted yeasts:-

a. Suspend a portion of each suspected colony type in a tube of sterile distilled water

b. Streak a loopful of the suspension on to a SAB agar.

c. Incubate the SAB plate at 30°c for 48 hours.

d. Check for purity. If the yeast is not pure, it must be restreaked. Note that some strains have both rough and smooth colony types in pure culture.

5. Mould Contamination

1. Colony isolation on yeast malt (YM) agar.

a. Suspend a small portion of the yeast mould colony in sterile water.

b. Streak a loopful of the suspension for colony isolation onto a plate of YM agar.

c. Incubate the YM plate at 30°c for 4-6days.

d. Check for purity. If not pure steps are necsssary (Step2)

2. Colony isolation in yeast malt broth.

a. Transfer a small portion of the yeast-mould isolate to a tube containing 10ml of YM broth.

b. Incubate the YM borth at 30°c for 48 hours. Remove a small portion of the sediment with a sterile capillary pipette by slipping the pipette along the edges to the tube to the bottom without disturbing the mycelial pellicle.

c. Streak the sediment for colony isolation on to plate of YM agar.

d. Incubate at 30°c for 4-7days and then prepare a direct mount. If the yeast is not pure further steps are necessary (step 3)

1. Colony isolation in shake culture.
   a. Transfer a small portion of the yeast mould isolate to a 250 ml Erlenmeyer flask containing approximately 100 ml. of YM broth.
   b. Place the flask on a rotary shaker and incubate at 30°c for 4-6 days.
   c. Carefully remove a small amount of the sediment with a sterile capillary pipettee. Ensure that the balls of mycelium are not removed by accident.
   d. Streak the sediment for colony isolation on to a plate of YM agar.
   e. Incubate at 30°c for 4-7 days and then prepare a direct mount using a small portion of each yeast colony to be identified. If the yeast is not pure, consult the supervisor.

Identification Scheme

To identify yeasts, first examine the colony color, shape and texture. If the colony is black to brown in color or moist mycelial in texture, prepare a direct mount. Using the flow chart, determine the genus based on microscopic morphology. If the colony is pink to red, streak the colony out for islaotion. After incubation, check for the presence of satellite colonies. Examine microscopically for forcibly discharged conidia. Using the flow chart, determine the genus based on morphology. Assimilations will be necessary to determine the species of Rhodotorula. If the colony is white or cream color, perform a germ tube test. If the germ tube test is positive, the yeast can be identified as Candida albicans. If germ tubes are absent, confirm the purity of the yeast isolate and inoculate an Vitek YBC or an AP120C strip and a Dalmau plate. No identification will be performed on yeast recovered from respiratory sites except if the yeast resembles Cryptococcus neoformans or if the identification is requsted by the physician to the director. For patient care, all germ tube negative yeast from respiratory sites will be reported as yeast not Candida albicans or Cryptococcus sp. Most of the commonly receoved ylasts can be identified to the species level using the morphology on corn meal agar and assimilation results (See Dalmau morphology flow chart) Rarely, it is hydrolysis cycloheximide resistance or ascospore induction. These procedures are available to aid in yeast identification. Problem identifications should be brought to the attention of the supervisor, Identifications using a commercial method will be carried out by using materials from the clinial Microbiology Laboratory which has been checked for QC.

Identification Procedures

A. Direct Mounts-Direct mounts are made in order to study yeast morphology microscopically and to determine purity of the isolates.

B. Lactophenol Mount:-
   (1) Place a small drop of lactophenol (LP) on a clean glass microscope slide.
   (2) Remove small portion of the yeast colony and place it into the drop of LP and suspend the cells.
   (3) Place a clean cover glass over the suspension and observe microscopically.
   (4) Seal edges of the cover glass with finger nail polish to temporarily preserve mount.

IDENTIFICATION YEASTS

A scosporogenous Yeasts

Some yeasts will readily form ascospores on primary isolation medium, whereas other require special media. These form a true myceluim or pseudomycelium. Vegetative reproduction is for budding, binary fission or arthrospores.

1. Vegatative reproduction by fission, with a true mycelium and arthrospores formed:
   No budding occurs————1 Endomyces.
   2 Schizosaccharomyces
   Vegetative reproduction by multilateral budding and sometimes fission—2
   Vegetative reproduction by bipolar budding includes 3-Hanseniaspora
2. True mycelium formed as well as budding cell-4 Hyphopichia
   No true mycelium but a pseudomycelium and / or loose collections of budding cells may be formed —— 3 and 5 Hansenula.

3. Dry dull pellicle develops rapidly in malt extract broth —— 4.
   Dry pellicle develops very slowly if at all in malt extract broth ——— 5
4. Ellipsodial to long cylindrical cells.——— 6 Pichia
   Round or short ovoid cells——7 Debaryomyces
5. Glucose always strongly fermented———8 Saccharomyces Kluyveromyces and Zygosaccharomyces
   Glucose not fermentled or only weakly so—— Debaryomyces

1. ENDOMYCES

It forms a true mycelium, and its ascospores are spherical, ovoid or hat shaped. Themycelium breaks up into arthrospores which are either cylinderical with rounded ends or ovoid (See Fig.) Vegetative reproduction is by fission. The equivalent genus in the Fungi imperfecti is Geotrichum. The species Endomyces lactis (also known as Geotrichum Candidum) is commonly found on dairy products and is therefore called 'dairy mould.' The colonies are white in colour and are yeast like and butyrous, particularly when older.

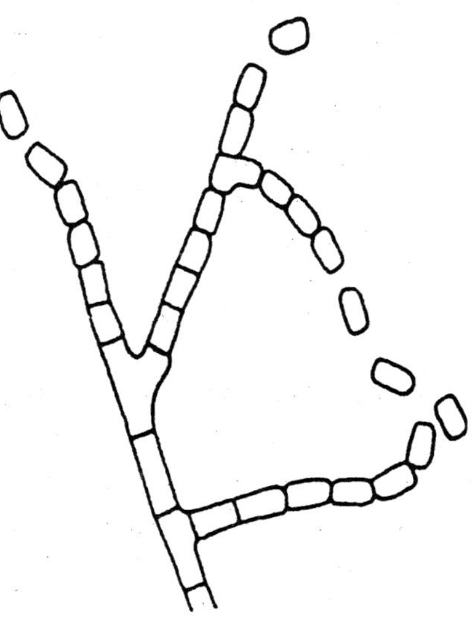

Fig. Endomyces

2. **SCHIZOSACCHAROMYCES :** These are found in food products such as fruit juices, dried fruits, honey, molasses etc.

The cells are cylindrical ovoid or spherical, but a mycelium that breaks up into arthospores may be formed. Vegetative reproduction is by fission. Eight or four ascospores are formed per ascus. Schizosaccharomyces may have a similar appearance to Endomyces but in the former asci are formed after the fusion of two arhrospores, whereas in Endomyces asci develop in the mycelium at branches of the hyphae. Schizosaccharomyces ferment carbohydrates. It cannot use ethanol as a sole carbon source.

3. **HANSENIASPORA:** The equivalent genus in the Fungi imperfecti's Kloeckera. The cells are ovid or lemon-shaped. A pseudomy celium is rarely formed. Vegetative reproduction is by budding at both poles. Ascopores (one to four per ascus) are spherical, becoming hat shaped or saturn-shaped. Hanseniapora can ferment carbohydratcs. It can not use ethanol as the sole carbon source.

4. **HYPHOPICHIA:** Budding cells are typical yeast and may from either a true mycelium that gives rise to blastospores or a pseudomycelium, in these respects resembling candida. Ascospores may be one of a variety of shapes depending on species, including spherical or ovoid. A pellicle is usually formed in liquid media. Hyphoichia attacks sugars oxidatively and does slow fermentation. It grows only slightly with ethanol as the sole carbon source. Members of this genus have been isolated from bread, fruit and vegetables.

5. **HANSENULA:** This genus reproduces by multi bacterial budding and ascospores Pseudomycelia ae frequently formed, particularly by species with cylindrically shaped cells. These yeasts assimilate nitrate. Ascospores (One to four per ascus) are spherical, hat shaped or saturn-shaped. Glucose is fermented vigorously, other carbohydrates being fermented according to species found in grains, fruit, shrimp and brines of pickles.

Some species and strains are useful. According to Batra ad Millner (1974) certain species are used in India to produce Kanji and rice wine. A strain of Hansenula grows on methanol to produce single cell protein.

6.**PICHIA:** The cells are ovoid, ellipsoidal or cylinderical, reproduction is by multilateral budding with most species forming a pseudomycelium. On solid media colonies are white, or creamish and usually wrinkled. Ascospores(One to four per ascus) are variously shaped, being spherical, hemispherical, hat-shaped, or saturn shaped. Carbohydrates are not usually fermented, or not weakly so. Many species can use ethanol as a sole carbon source. Pichia is an important contaminant of beers, pickled products, fruit and wine. They can occur as films on brined foods.

7. **DEBARYOMCES:** The vegetative cells are often spherical. The sexual ascopores, produced by Configuration of the mother

cell and bud are spherical or oral. These are usually one of two spores per ascus. Carbohydrates are not fermented or only weakly so, some species can use ethanol as sole carbon source. Debaryomyces has a high salt tolerance and can frequently be isolated from pickled and salted foods.

8. **SACCHAROMYCES, KLUTVEROMYCES AND ZYGOSACCHARMOYCES:** The group is considered to be a rather hetrogenous group of organisms. The cells are spherical, ovoid or longer cells with rounded ends. Vegetative reproduction is by multilateral budding. Pseudomycelium may be formed but a true mycelium is absent, on agar the colonies are generally white or cream with a typical yeast odor, In liquid media, sediment is formed, sometime with ring growth at the air-liquid-glass junction, a pellicle is not formed

Ascospores ((one to four per ascus) are usually spherical or ovoid. In kluyveromyces the mature asci-rapture to release the ascospores. In the case of saccharomyces, rapture of asci does not occur on maturity. All species have a vigorous fermentation(see table) Species are defined mainly on the basis of carbohydrate fermentation and carbon source utilization. Although morphological differences species can be seen. There is usually too great a morphological variation within a species for characteristics such as cell shape to be used with any degree of reliability. S Cerevisiae is the top yeast used for production of alcohol(brewing) and carbondioxide (baking). Besides these important uses, species and strains have been used to remove glucose from egg white prior to drying, to remove the mucilage layer from coffee beans during processing. The strains of Kluyveromyces marxianus previously included in K.(Sccharomyces)fragilis and K(Saccharomyces) lactis are found in certain fermented milk''. The organisms in this genus are associated with variety of food can cause spoilage in fruit, fruit products, sugar syrup, honey, vinegar, wine and salad dressings etc. At present time, no pathogenicity has been established for species of saccharmyces.

Differentiation of Common species of Saccharomyces Klutveromyces and Zygosaccharomyces.

Key to the identification of Yeasts Dalmau Morphology Flow Chart

Ascospores present							Hansenula Pichia Saccharomyces
Ballistoconidia present							Sporobolomyces and similar yeasts
Basidiospores present							*Filobasidiella* *Filobasidium* *Leucosporidium*
Hyphae, pseudohyphae, or both present	chlamydospores	Present					*Cadida albicans*
		Absent	arthroconidia	Present	blastoconidia	Present	*Trichosporon*
						Absent	*Geotrichum*
				Absent			*Candida* *Saccharomyces*
Sporangia present							*Prototheca* *Sarcinosporon* *Fissuricella* or similar fungi
Yeast only							*Cryptococcus* *Rhodotorula* *Torulopsis* *Candida guilliermondii* or an ascomycetous yeast not producing ascospores, i.e., *Saccharomyces*

YEAST MORPHOLOGY FLOW CHART

Brown to black colony			Direct Mount		Yeast only	*Phaeococcomyces*
					Yeast and hyphae	*Aureobasidium* *Exophiala* *Wangiella*
Pink to red colony	Satellite colonies	Present	Forcibly discharged conidia present			*Sporobolomyces*
		Absent			Purify	*Rhodotorula*
White colony			Germ Tube Test	Positive		*Candida albicans*
				Negative	Purify	Dalmau plate YBC-Vitek
Moist mycelial colony			Direct Mount	Arthroconidia		*Geotrichum* *Trichosporon*
				Other types of conidia		hypohomycete

* Young colonies of *Cryptococcus neoformans* are often indistringuishable from young colonies of *Candida*. Therefore, technologists performing the germ tube test must be alert to yeasts that have a microscopic morphology suggestive of *C. neoformans*.

Table : Differentiationof common species of *Saccharomyces, Khuyveromyces* and *Zygosaccharomyces*

Organism	Glucose fermentation	Maltose fermentation	Galactose fermentation	Sucrose fermentation	Lactose fermentation	Fructose fermentation	Utilization of ethanol as sole carbon source	Cycloheximide resistance
S. cerevisiae	+	V	V	V	–	+	V	–
S. uvarum (S. carlsbergensis)	+	V	V	+	–	+	–	–
*Kluyveromyces marxianus**	V	V	+	V	V	+	+	+
Zygosaccharomyces rouxii	+	V	–	–	–	+	V	–
Z. bisporus	+		–	–	–	–	+	–

* Synonyms include *K. fragillis* and *K. lactis.*

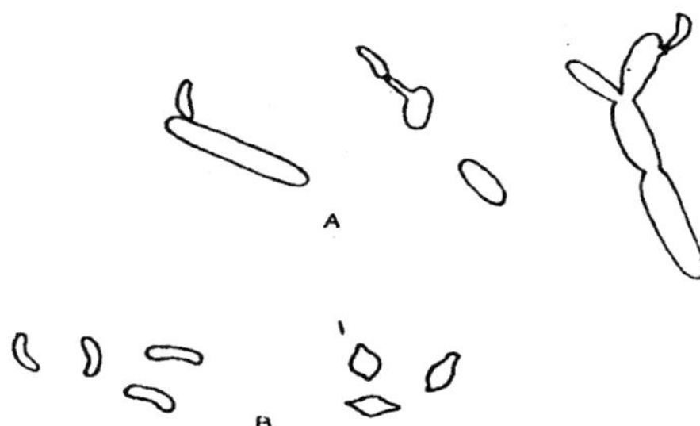

Fig. I : Yeasts of different shapes

SPOROBOLOMY CETACEAE: Ballistospores are formed and discharged forcibly into the air when ripe, and their may give rise to "mirror images' of the colonies on the lid of the petri dish or the lower slide when these yeasts are grown in plate cultures or two-slide cultures. The ballistospores may be spherical, ovoid, kidney, shaped, bean shaped or sickle (see Fig I)

In liquid media growth is predominently on the surface, although slight sediment may be formed.Neither sporobolomyces nor Bullera can ferment carbohydrates. Colonies red, pink or salmon-pink in colour, although sometimes very pale or even cream. A pseudomycelium or true mycelium may be formed. Ballistospores kidney or sickled shaped is Sporobolomyces, Colonies colourless, white. Cream or pink. A pseudomycelium or true mycelium is never formed. Ballistospores spherical or ovoid of Bullera.

ASPOROGENOUS YEASTS

TRICHOSPORON: The cells of Trichosporon species are of various shapes. Buds are multilateral. True mycelium and arthrospores are formed. The presence of arthospores differentiates this genera from candida. Pseudomycelium may be developed by the budding cells. These organism are found in various foods such as fresh shrimp, Crab., Beef, butter, cheese, fruit, fruit juice and rice.

CANDIDA

All species form pseudomycelium and by fussion, some form true mycelium and chlamdospores. They can produce alchohol by fermentation. These organisms have been involved with spoilage of various food such as fresh fruits, vegetables, dairy products, brine and alcholic beverages, pickled products.

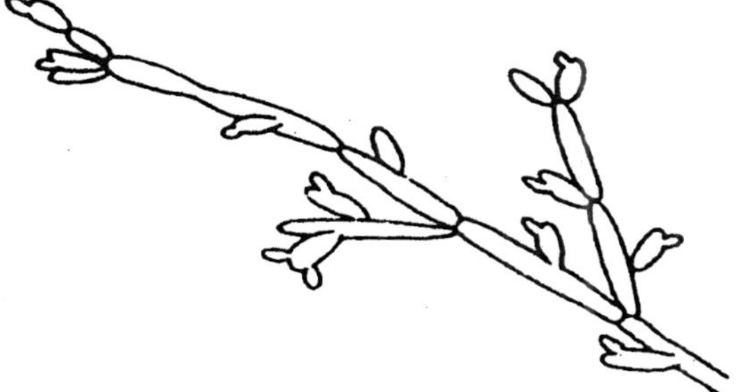

Candida showing blastospore production (see figure)

RHODOTORULA: The cells reproduced by multilateral budding. There may be a rudimentary pseudomicelium. The cells are brightly coloured red, orange or yellow. They are involved in the spoilage of a wide range of food products. Members of this genus do not ferment carbohydrates. They cause slime surface on cottage cheese, spoilage of area, butter, refrigerated beef and sweetened condensed milk.

BRETTANOMYCES: The cells are spherical or ovoid cells pointed at one end (Ogive` shaped) A pseudomycelium may be formed. On solid media, colonies are white to yellowish, and may be glistering, moist and smooth, or dull and wrinkled. Carbohydrates can be fermented by some species. Acetic acid can be produced aerobically from glucose. Ethanol can be utilized. This organism can cause spoilage in beer, cider and wine.

TORULOPSIS: The cells are spherical or ovoid and reproduce by multiplier budding. A rudimentary pseudomycelium is rarely formed. It is rarely capsulate. On solid media, colonies are off-white, cream, yellowish or brownish, smooth and may be shiny or dull.

Carbohydrates may be fermented. Some species utilize ethanol as a sole carbon source. They can cause spoilage in pickled products and food with high sugar concentration.

MOULDS: Moulds are part of large group of micro organisms called fungi. The fungi are ubiquitous being found everywhere, soil, air, water and decaying organic matter are prime source. They differ from bacteria by their complex structure and greater size, The fungi may be multicellure or unicellur.

The fundamental structural units of moulds are filaments tubes called hyphae. By formation of cross walls or septa some hyphae form chains of cells that are septate, others may not form septa and the hyphae are nonseptate. The septa have pores that allow the movement of cytoplasm from one cell to another. As the hyphae elongate, they interwine. A mass of these interwined branched hyphae is called mycelium. Part of the mycelium grows into the substrate and absorb foods. This is known a Vegetative mycelium. The mycelium that remarks in the air above the substrate and bears spores is called the aerial or reproductive mycelium. Moulds can reproduce sexually or asexually or by both systems.

They are several types of spores formed by the fungi. The asexually fungi produce spores directly from or by the mycelium. These are called thallospores, conidiospores or sporangiospores. They are three types of thallospores, (1) Blastospores (2) Chlamydospores and (3) arthrospores. Sexual spores inculde Zygospores, ascopores and basidiospores.

IDENTIFICATION OF MOULDS: Identification of moulds is based almost entirely on the structures bearing spores and on the spores themseleves. These are differentiated by cultural examination and precaution is taken for distortion of sporing structures.

The media on which mould is grown has peromed infulence on the colonial appearance and on the developments of aerial sporing structures. Chlamydospores, Sclerotiae etc. The low pH media (pH4.5) inhabitated the growth of Bacteria. In some medias antibiotics are used to inhabit the growth of bacteria and pH is near 7.0.

Microscopic methods are used to enumerate mould filaments in Tomato products and other canned fruits and vegetables. Slide culture methods are used to prevent spores being detached, the method has been recommended for yeast as described earlier.

SLIDE CULTURE METHOD: A Petridish containing filter paper disc in the base, on which is a U-shaped glass rod carrying two clean microscope slides. Sterilize it. After sterilization aseptically add a few millilitres of a sterile 20% solution of glycerol to the filter paper to keep the slide culture from drying up during incubation. Take another petridish and pour 15mls of medium and allow it to set. After setting the medium, cut blocks of 1cm^2 and mount them on the microscope slides. Inoculate the four cut surfaces of the each block with the mould to be examined and place a cover slip over the block. Incubate the slide cultures in the petridishes with the lids in plate. The slide cultures may be examined from time to time under low or high power dry objectives. The coverslips help to prevent contamination at this stage. When mould growth appears, a wet preparation can be made, as the mould tends to adhere to coverslip and slide. Remove the cover slips from the agar block and mount on fresh slides using Lactophenol-Cotton blue stain. In addition the agar blocks can be removed from the original slides and these slides also examined with the staining mountant and fresh coverslips.

IDENTIFICATION OF MOULD BASED ON MORPHOLOGY OF SPORING STRUCTURES

THAMNIDIUM: The mycelium formed in loose and cotton wool like, up to 1cm or more in height, light grey at first, darkening as a result of the production of sporangia and sporangioles. Zygospores are produced at 6-7°c but not at 20°c.

The hyphae are non septate and of large diameter. The long main sporangiophores carry whorls of short, dichotomously-branched, sporangiophores. Each long spo-rangiophres terminates in a large globose sporangium containing numerous spores, Whereas the short sporangio-phores terminate in small sporangiophores each containing one to four spores.

The mould is found on meat as well as on soil and animal carcasses kept at refreigerated temperature.

Mucor

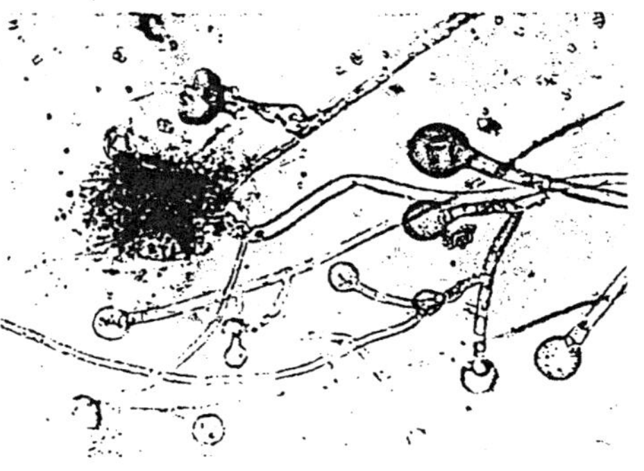

The hyphae are non-septate and of large diameter, the younger the hyphae, the smaller the diameter. The sporangrophores are erect and may be unbranched or branched. Each sporangiephore or branch bearing a single globose sporangium containing a large number of spherical or ellipsoidal spores. The sporangiophores never arise from nodes on stolons, and rhizoids are absent from the mycelium. Stoloniferous growth does not occur in mucor which differeniate this genus from Rhizopus.

The growth on a solid medium gives a loose cotton wool-like aerial mycelium. First white or grey in colour, latter becoming darker as sporangia are product. A few species do not produce a white or grey myceilum at first, being yellow, orange or bluish in colour. In addition, a few species produce yellow. Mucor occurs in soil, fruits vegetables, stored grains and other foods. M.Pusillus produces an extracellur protease which has milk clotting activity.

Rhizopus

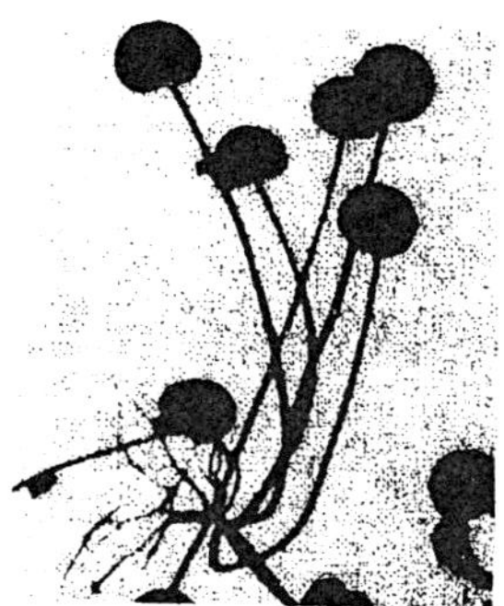

The mould grows very rapidly. It is white and cottony when young and quickly turns brown to black. The spores appear as little black dots. This mould grows well on bread, as well as fruits and vegetables left on kitchen counters. To reduce mold on fruits and vegetables, store them in the refrigerator until ready to be used. Fruits and vegetables should be washed well or peeled before eating to avoid inadvertent mold ingestion.. The persons suffering or have suffered with bone marrow transplant, leukemia, and diabetes, this mould is a source of severe and sometimes fatal infection. In people with normal immune system, these moulds are not infectious, but it is possible to have allergic reactions to them. Rhizopus produces high yields of nurmeric acid from fermentable sugars. Species are used in the production of fermented foods such as Tempeh. It is present indoors and outdoors.

Syncephalastrum:

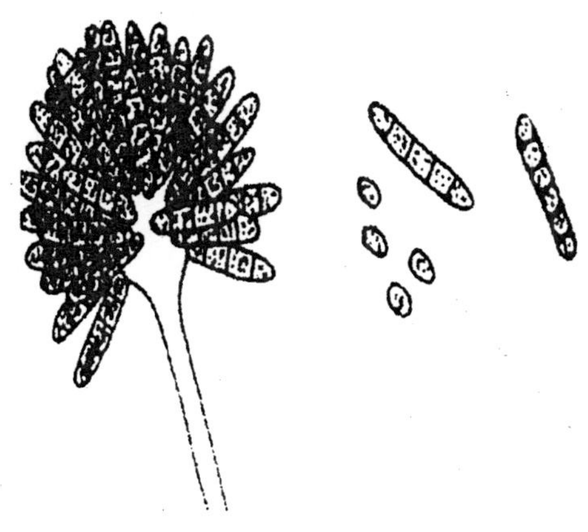

It possesses a non-septate mycelium of hyphae of large diameter. The sporangiophors have branches which are not cut off by septa, have a terminal swelling. Each terminal swelling bears a number of cylinderical sporangia, each of which contains a chain of spore The colonies are grey to black.

Sporotrichum:

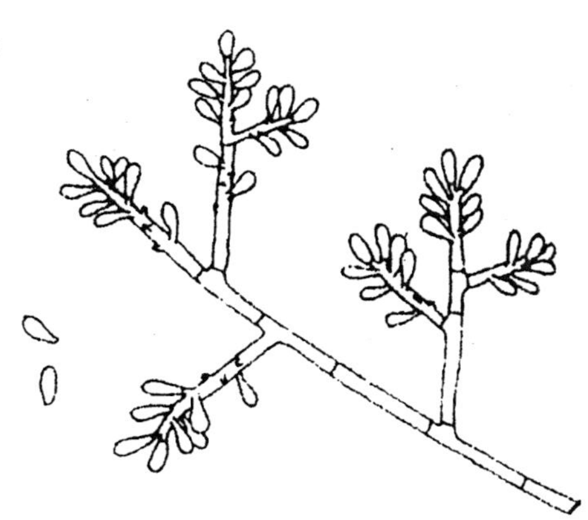

The Conidia are born singly on very short projection which arise near the ends of the hyphae and do not occur in chain. The Conidia may be spherical or ovoid, colourless or lightly coloured and transparent.

The growth of colonies on medium usually white but may be yellow, grey, pink . red or green. Sporotrichosis caused by S.Schenchic is a chronic, progressive inflection of the skin and subcutaneous tissues. S.Carnis grows at low temperature (–5 to 8°C) and can cause a defect called "white spot" of refrigerated meat.

Bortytris: The asexual form of Botryris reproduces by Conidia. The conidiaphores develop from a sclerotium, and are irregularly branched. The Conidia occur on short sterigmata. The positions and number of sterigmata cause the conidia to appear in graph-like

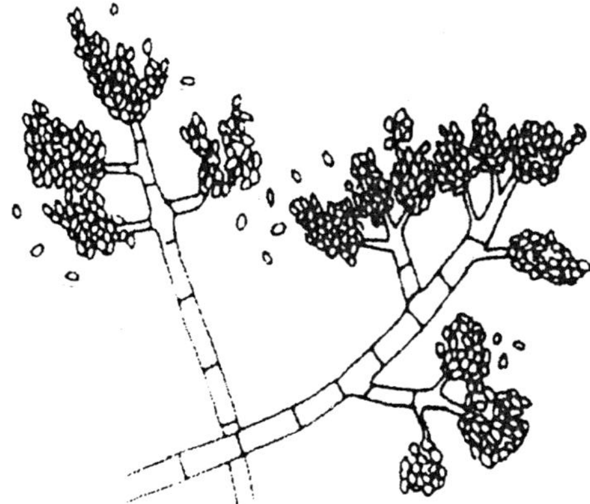

clusters. Some of the species in this genus are imperfect form of the sexual form. Botryotinia Becnera is the common species of Botrytis Colonies are frequently white and fluffy at first. One media with a low C:N ratio, Conidia are formed. Media with high and low C:N ratio can be made by using malt extract agar with modifications to the concentrations of malt extract and peptone. Reduction in the amount of peptone provides a medium with a high C:N ratio, whereas reducing the amount of malt extract and increasing the amount of peptone produces a medium with a low C:N ratio. It is the gray mold of various plant products especially lettuce, tomato, strawberry, rasberry and grape. It can be classed as field mold since it is a common soil contaminant and attack fruits and vegeables in the field. It enters fruits such as tomatos through cracks and injured areas.

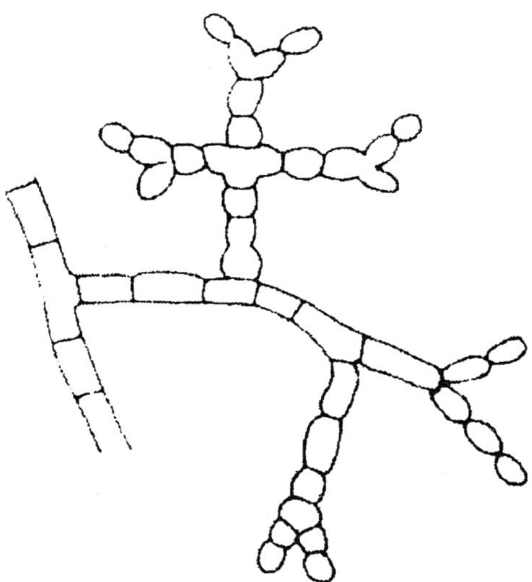

Monilla : A Conidia occur in branched chains and chains are due to budding of the Conidia. . The mycelium as it ages tends to form arthrospores. Monilla Sitophilla is the commonest and most well known species. It grows best on organic media and colonies are at first white and very loose and fluffy but rapidly become pale pink to red in colour due to production of masses of coloured Conidia. The growth rate of Conidia is very high. M.Sitophila can therefore easily become a containment in microbiological laboratories and incubators and also grow on bacterological media and at temperatures up to and including 37°c. It spoils bakery products especially wrapped, sliced bread, causing a pink fluffy growth which is extremely characteristic and easily to recognize.

WALLEMIA: Conidiophore are usually non pigmented and produced erect from the substrate mycelium. The Conidiophore terminates in a hypha which is non-septate at first, but develop septa, and produces endospores or arthrospores which are usually pigmented.

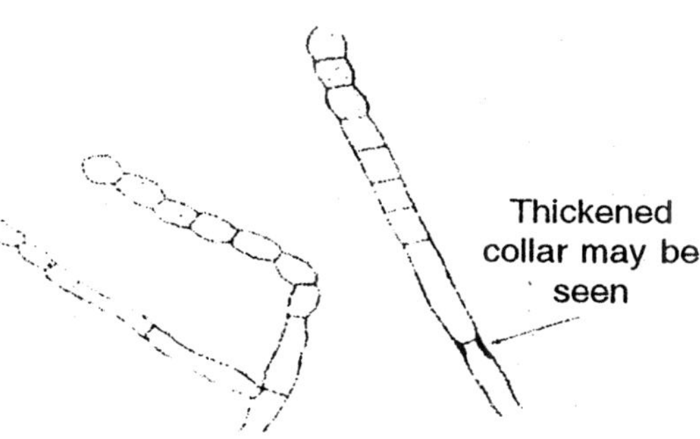

Wallemia Sebi may be found in bakery goods and in foods with high sugar or salt content. It produces brown colonies in the media.

ASPERGILLUS : The mould thrives in damp warm conditions. These are over 100 species in this genus. The most common is Aspergillus farmigatus, initially a yellow-white mould, which rapidly changes to green and finally black. Other species of Aspergillus have different colours. Asperigillus flavus are toxigenic species. These attack grains, spices and nut produce Aflatoxin B_1+B2. They are capable of spoiling of food of high concentration like Jams and Plum puddings.

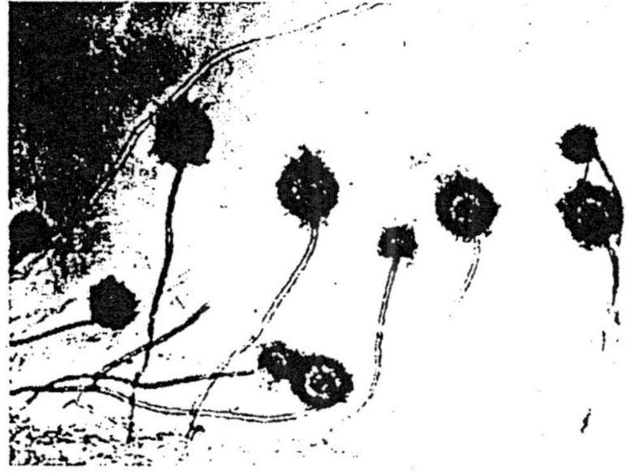

The organisms are a source of amylase and pectinolytic enzyme. A proteolytic enzyme of Aspergillus is able to clot milk and a substitute for rennet in cheese making.

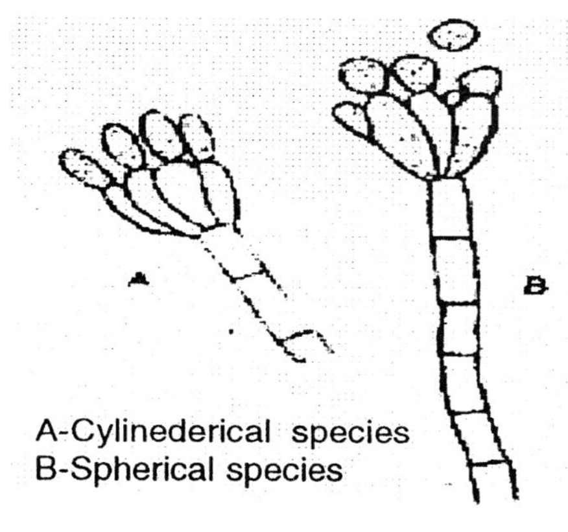

A-Cylinederical species
B-Spherical species

STACHYBOTRYS: The hyphae is transparent the conidiophores become dark as they mature and the spores are usually dark in colour. The phialides which bear the Conidia are characteristically swollen, and borne as apicalcluster of only three to seven on each conidiophore. The spores may be spherical, ovoid or cylinederical.

Some species produce spores that are dark in colour, but cause the colonies to appear pink or salmon. The common species of strachybotrys are cellulolytic

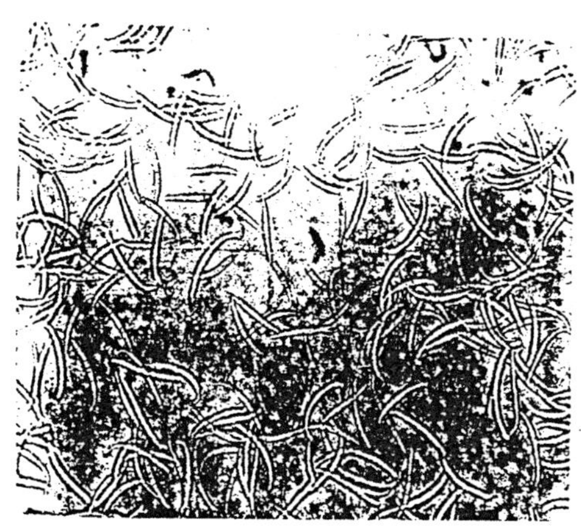

FUSARUIM : It is a common soil fungus, which sporulates in warm, wet, weather and is widely found on grasses and other plants. The coindia pro-duced by members of this genus have various shapes such as spindle or sickle-shaped. The formation of coindia and the colonial characteristics depend largely on the medium used, but no general recommendations can be made concernings the choice of the medium as the correct cultural conditions for the formation of the Macrocoindia vary according to the species.

However Potatoes dextrose agar and Potatoes carrot dextrose agar may be tried with and without glucose. The colonies may be fluffy and spreading. Colonies of many species of Fusaruim have been seen as White, Rose, Pink, Rose red, Brown, Purple and Yellow.

The fungus can cause plant diseases and a major parasite of rice, sugar, cane, sorghum, and especially maize grains. It occurs regularly on banna roots and other fruits and vegetables such as tomatoes and watermelon. It can cause allergic fungal sinust's and is resistant to all antifungal agents.

Chlamydospores are frequently formed in the mycelium and even in the Coindia. F.moniliforme causes a disease of rice which led to the discover of gibberllic acid, a plant growth stimulant. The fusaria are widespread in nature, being found in Soil, decaying material and food. Some years fusaira have caused a 50% loss of wheat and other crops in Japan. F.Solanie causes a decay of Potatoes called powdery rot, dry rot or white rot. The fusaria produce mycotoxins which effect various animals and possibly human.

PENICILLIUM: It is one of the most dominant, and important house moulds, the indoor mould can be readily seen on the stale bread, citrus fruits and apples. It is frequently found in wine cellars The antibiotic Penicillin is produced from only a few strains of Penicillium, which occur really in normal exposure.

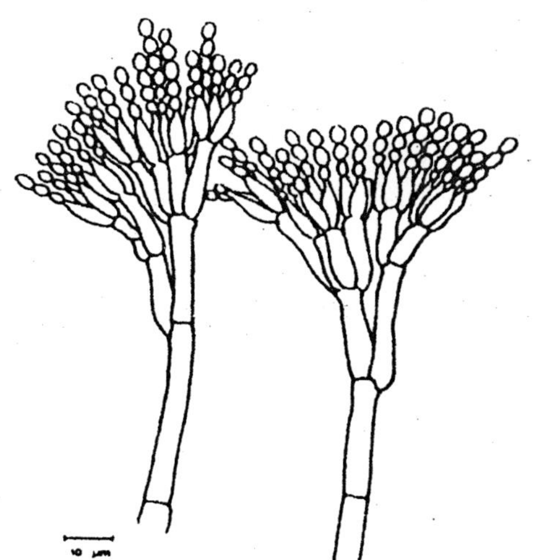

There are many species of Penicillium. They are closely related to the aspergilli. The Pencillia are characterized by branching of the conidiphore to form a brush like conidial head. This moulds has a velvety colony that has blue green centers with pale to bright yellow and yellow exu-date. It has a fruity odor like apples or pineapples.

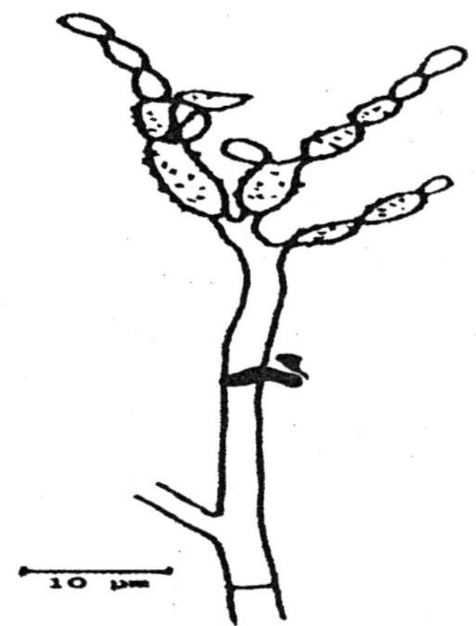

CLADOSPORIUM : Cladosporium is also known as Hormoden-drum. The conidia are one of two-celled and produced in branched chains on the Conidiophores. The conidia reproduces the branches by budding. The colonies are fairly thick, dark olive green with a velvet-like folded surface. They can grow on connective tissue or fat covering of meat kept in cold storage. The growth results black spots on the meat. In nature, the spores are released in great numbers after rains. Cladosporium is the most abundent mold spore found in air samples and is second only to Alternaria in the frequency with which it causes allergy smptoms. It is rarely or never a cause of allergic fungal sinusitis.

ALTERNARIA: This is an outdoor mould, which frequently grows in food stuffs and soils. The black spot seen on tomatoes are usually Alternaria. The mould appears when the weather is warm and is often found on condensed window frames.

The Conidia are pear or club-shaped, multicellular with both crosswalls and longitudinal septa, ad produced in chains. The mycelium and mature Conidia are darkly coloured. The colonies may be dark green, greenish, black, brown or yellowish brown. It is also involved in the development of rancid or off flavours in dairy products attacking injured or weakened tissues. Due to darkening of the tissues, the defect is called black rot. These organisms also are involved in the development of rancid or off flavours in dairy products.

Microbiological Examination of Foods

Introduction

The responsibility of Food Microbiologist in the food industry in quality control of incoming of raw materials, quality assurance/ control of production. Maintenanace of hygienic atmosphere, use of proper detergent-disinfectants and proper hygienic code of practice in day to day production.

It is not possible for any food company who can protect the production up to zero tolerance on food spoilage organisms. The only way is proper control by food handlers and production staff. The procedures prescribed by ICMSF must be followed:

It is also important to know the historical facts of food poisoning out break by a particular pathogens present in a particular food. The known records of involvement give us the idea which tests should be performed on food.

The examination of complex food products and their constituents for the presence of potential spoilage or food poisoning organisms also require care. For example certain species may give very high viable count. Black pepper normally contains a large number of aerobic spore bearers. The introduction of pepper into a food product at end of its preparation may thereore add large numbers of Bascillus, which may under certain circumstances, be able to multiply sufficiently to cause spoilage or B.cereus food poisoning. Thus each portion of complex food product require correct selection of tests and counts.

METHOD OF SAMPLING

Sampling of food stuff is very important to arrive at giving the verdict. Various procedures have been followed by different countries. The international commission on Microbiological specification for foods (ICMSF, 1986) discusses in detail the criteria for determining the stringency of sampling that may be required for different kind of food. The International Standard Organisation (ISO 948) also lay down specifications for sampling. The general methods followed for different kind of foods are described below:

Liquid Samples

The liquid food (e.g. Milk, Ice-cream mix, sugar. Syrups, juices)should be mixed thoroughly up and down before a sample is taken. The sample about 100ml to 500ml should be drawn with the help of sterile spoon and kept in sterile container for Laboratory checkup.

Solid Samples

The material may be drawn with the help of sterile spoons, Scalpels and a sharp steel long knife of canal shape depending on the type of material to be sampled. The food like Milk powder, flour is capable of sufficient mixing and quantity of 100g is drawn from different sides of the container. When the Bulk quantity is to be tested, number of samples are drawn, mixed thoroughly in the Laboratory, before smaller samples are removed for testing. Canned Products the sample should be drawn from different cans, minimum 2 percent should be drawn in the Laboratory from sampled Cans under sterilized conditions(Sterilized Chamber). Frozen foods samples can be drawn by electrical drill fitted with a bore-extracting kit, the place is cut deeply. No need of thawing.

Sampling Surfaces

The sample in the form of slice from the upper layer is removed and put in the medium. Alternatively surfaces may be examined by transferring the microorganisms from sample to microbiological medium with the aid of a supposedly inert carrier that neither causes death nor allows multiplication of the microorganisms. Such carriers include rinses, swabs and adhesive tape. For long time the microorganisms may die on the carrier or multiply. It is better to inoculate into suitable media as early as possible.

An other method is to transfer the organisms directly the sample surface to the surface of the medium by impressing the one upon the other.

Surface Slices

Remove very thin slices of the superficial layers of the food using. Sterile scalpels and forceps. Homogenize the slices in a suitable diluent to obtain an initial 10^{-1} dilution.

Rinses and Washes

Rinse or wash the food (one part of weight in sterile diluent (ten parts by weight) and consider the washings to be initial 10^{-1} dilution. This procedure is applicable to food such as Pork, Ham, Sausages dried fruits, vegetables and salad vegetables. It is to be noted that the microorganisms on the surface of the food will not be detached by merely shaking in the diluent so it is advisable also to obtain samples from which dilution can be prepared by comminution. The extent of shaking and consequent removal of microorganisms should be standardized. The reporting represents the bacteria on the surface only.

Cotton Wool Swabs

Cotton-Wool Swabs are prepared from non-absorbent Cotton wool wound to a length of 4cm and thickness of 1-1.5cm on wooden sticks. They should be placed in alloy tubes which are then plugged and sterilized.

Procedure

Moister the swab with sterile quarter-strength Ringer's solution and rub firmly over the surface being examined, using parallel strokes with rotation of the swab. Repeat this process using parallel strokes at right angles to the first set. Care must be taken that the whole of predetermined area is swabbed. Replace the swab in the tube. To prepare counts, add 10ml of quarter-strength Ringer's solution. Agitate the swab up and down in the tube ten times to assist the rinsing of the bacteria from the swab. Prepare plate counts from 1-ml amounts of the swab washings and from dilutions prepared as required. Count should be recorded as the number per square centimeter of surface swabbed.

Adhesive tape (Self Adhesive Lables)

The adhesive strip or label is removed from the mount, pressed against the surface to be examined, pulled off immediately and replaced on its mount. In the laboratory, the strip is removed from the mount, pressed against the surface of an appropriate culture medium, and then removed and discarded. The adhesive tapes are used for investigation on the microbial flora of human skin, food surface and equipment surfaces.

Contect Slides

Press a sterile glass slide against the food sample to be examined. Transport the slide back to the laboratory. It may be examined microscopically after fixing and staining (e.g. by Gram's method). Alternatively the slides may be impressed on to poured plates of media, to achieve some transfer of the organisms to the agar surface. After removal of the slides (using Sterile forceps) the plates then may be Incubated. This method is useful for rapidly determination of Microflora especially on raw meat, poultry and soft cheese.

Alginate Swabs

These swabs are prepared from Calcium Alginate wool. About 50 mg of wool is recommended for preparing the swabs and wound, on wooden sticks giving a swab 1.5cm long by 7mm in diameter, moistened very slightly in quarter strength Ringer's solution. These are kept in 10x1 cm metal tube and sterilized by autoclave at 121°c for 15min.

Method of Use

Rub the area of the surface to be examined in parallel strokes, with slow rotation of the swab. Swab the same surface second time, using parallel strokes at right angles to the first set. When laboratory examination is required, break of the swab aseptically into a sterilized screw capped bottle containing 10 ml of a sterile 1% solution of calgon (Sodium hexametaphosphate) in a quarter strength Ringer's solution. The contents are shaken, of this causes the alginate wool to disperse and dissolve giving a suspension of all the bacteria present on the Swab. Take 1ml amount in a sterilized petri-dish and add tryptone soya agar (as described in plate count method) Counts should be recorded as the number per square centimeter of the surface swabbed.

Sampling for Count of Anaerobic Bacteria

For anaerobic bacteria examination, as an example a small quantity of sample from deep tissues of meat containing little oxygen is to be drawn and should not exposed to normal atmospheric concentration of oxygen. When swabs are used for taking the sample a suitable transport medium (e.g. stuart's transport medium) which is capable of reduced conditions must be employed. Thus, if an alginate wool swab is used, it should not be reinserted into its original tube but placed in a bottle of stuart's transport medium. The swab may also be moistened with reinforced clostridia medium before use.

When the concentration of bacteria in food stuff is very low, it is difficult to use colony count techniques, it is possible to count by using multiple tube technique by employing successive dilutions as described in previous chapter of MPN methods. When desirable a large quantity (e.g.100, 25g and 10g) of food can be used and suitable dilutions are prepared.

An alternative approach, known as attributes sampling scheme, is to take a number of sample of the same size, and to determine whether any contain the organisms in question. If organisms are not detected it is then possible to calculate the maximum proportions of units of that size that will, with a given probability, contain at least one of the organisms in question.

When the food stuff is distributed already in discrete units as in the case with canned foods, it is preferably to adopt this unit as the basis of each sample.

The attributes sampling schemes are readily applied to problems such as determination of the proportions of cans of food, or packets of milk that contain spoilage organisms. It is not so obvious that attributes sampling scheme can be applied TO FOOD STUFFS THAT ARE NOT DISTRIBUTED IN A LARGE NUMBER OF DISCRETE UNITS (E.G. A Vat of Ice-cream mix) and also to foods in which the unit size is inconveniently large (e.g. a large can of frozen egg album)

In the above cases it is assumed that the distribution of the organisms is random throughout the batch, and that the batch of food is made up of units of a given size (e.g. 10 or 25g). A number 'n' of such units is examined for the presence of the organisms being sought. Then the proportion, in the entire back of positive units (i.e. units containing at least one organism) that may be detected with a given probability by at least one positive unit being found amongst the n units examined is given by:

$$d = 100\left(1 - \sqrt{1-p}\right)$$

This is derived from the formula $p = 1 - (1 - d/100)^n$ which shows the probability of at least one defective unit being in the sample of 'n' units taken, when d is the percentage defective in the entire batch. This relationship is true provided that is a small fraction (less than one-quarter) of the entire batch.

For example we take six units each of 25g and examine then for the presence of bacteria. Then the percentage of positive units in the whole batch that will be detected with a 95% probability. by at least one of the six units being found positive is given by:

$$d = 100\left(1 - \sqrt[6]{1-0.95}\right)$$

Hence $$\frac{d}{100} = 1 - \sqrt[6]{1-0.95}$$

or $$1 - \frac{d}{100} = \sqrt[6]{0.05}$$

Take log

$$\log\left(1 - \frac{d}{100}\right) = \frac{1}{6}(\log 0.05) = \frac{1}{6}(\bar{2}.6990)$$

$$= \frac{1}{6}\ (-1.3010)$$

$$1 - \frac{d}{100} = \text{antilog}\ (-0.2168) = \text{antilog}\ (\bar{1}.7832)$$

$$= 0.6070$$

$$1 - \frac{d}{100} = 0.6070 \text{ or } \frac{d}{100} = 1 - 0.6070 = 0.393$$

That is if all the units are negative then p=0.95 the number of microorganisms in the batch is such as to give less than 39.3% of 25g units positive, in other wards there are fewer than 40 bacteria per 2.5kg in the long run)

If we wish to find the number of units that needs to be sampled to detect a given proportion of defective units, we can do so from

$$n = \frac{\log (1 - p)}{\log \left(1 - \frac{d}{100}\right)}$$

For example suppose we wish to apply a standard for a food, that with a 95% probability, the food stuff contains less than one salmonella per 500g. How many 25g units must be sampled, and found to be positive, in order to detect a level of contamination of one organisms per 500g? There are 20units of 25g in 500g. Thus the standard we wish to apply is equivalent to there being less than 5% of defective 25g units (i.e. less than one unit in any 20 units being positive in the batch as a whole), with a 95% probability. The number of units 'n' that needs to be taken is given by

$$n = \frac{\log (1 - 0.95)}{\log \left(1 - \frac{5}{100}\right)} = \frac{\log 0.05}{\log 0.95} = \frac{2.6990}{\text{-}1.9777}$$

It means 59 units for a sample is to be taken each of 25gm, all of which must be negative, in order to ensure that in the long run the food contains (with a 95% probability) less than one salmonella per 500g.

Use of attributes sampling plans in conjunction with colony or most probable number (MPN) counts:

The ICMSF (1974, 1986) recommended application of three class attributes plans to microbial counts in general and used such plans as the basis for its recommended microbiological standards for a wide range of food stuffs. In a three class plan, instead of yes-no decision of the two-class plan discussed above, a further class is distinguished, the three classes being acceptable, marginally acceptable and wholly unacceptable. The two counts relating to the two thresholds are conventionally indicated as m and M respectively.

Whereas any count above the limit for unacceptability will lead to rejection of the batch, a certain portion c, of counts in the marginally acceptable or rejection of a batch, a certain portion c, of counts in the marginally acceptable range is allowable. Thus the use of a three class attributes plan in conjunction with counts such as general viable counts, coliform counts, etc. will require the testing of multiple samples taken at random. One of the stated advantages of three-class plan is that the probability of acceptance or rejection of a batch is less affected by nature of the distribution of the organisms within the batch than is the case with two class plans.

Choice of samples on a Non-Random basis

Usually attempts are made to take samples randomly. The random table may be issued to draw the samples. The samples are drawn from production line at the beginning of production and also from the pipe line during production as well as towards the end of the run.

In case of a batch of food in store, samples on a random basis may be justified if the storage conditions are the same for the entire batch. Often, however, temperature and other gradients will occur across the stack of food.

In the microbiological examination of a food production to determine the effects of the elements of that process (e.g. as would be required in setting up and verifying a Hazard Analysis critical control point (HACCP), the samples taken will be carefully chosen in order to verify the correct identification of hazards, their location points, and the appropriateness of the critical control points.

TRANSPORTS AND STORAGE OF SAMPLES

Whenever possible, the original state of the sample should be maintained until the Laboratory tests are carried out. Frozen foods should be kept frozen by using solid carbon dioxide (with precautions to protect the frozen food samples against exposure to the gaseous carbon dioxide) or 'Cold packs' in conjunction with insulated containers for transport to the Laboratory, followed by storage in a deep freezer until tested.

Perishable but unfrozen food samples should not be frozen before testing, but, if storage for a short period is unavoidable. The samples should be chilled and kept refrigerated at about 4°c until tested. The refrigeration of samples for period of 3 days or more will result in the multiplication of any psychrotrophic micro-organisms present in the food and may cause the death of some mesophilies or thermophilies.

5

Preparation of Dilutions of the Food Homogenate

Methods of isolation and enumeration of the microorganisms present in all non-liquid foods usually require preliminary treatment of samples of food to release into a fluid medium those microorganisms which may be embedded within the food or within dried or gelatinous surface films. In the methods for food sampling it is suggested that initial 1:10 dilutions of the food samples be prepared from 10g amounts of the food, which is the normal practice in many Laboratories. Nevertheless, it is recommended that if possible 1:10 dilutions are prepared by suspending or homogenizing 50g, as this gives much more reliable and representative results. The most common practice is to use an electrically driven mixing device with cutting blade revolving at high speed known as blander. The dilution thus prepared is used for various determination or enumerative procedures. Standardization of this preliminary procedure is important. Excessive speed of the cutting blades or under prolonged use of the mixer may cause injury to microbial cells either mechanically or by the heat generated. Thus excessive mixing may result in a reduced *count* and on other hand insufficient mixing in contrast may not release embedded bacteria or may provide a heterogeneous distribution within the suspension, Limitation of the speed of cutting blades and of the duration of mixing time is necessary to provide optional results. A blender with 15000 to 20000 RPM is used for time not exceeding 2.5 minutes. Care should be taken to minimize risk from aerosols when pathogens may be present in the test food. All mechanicals blenders create aerosole.

Now stomachers are used for blending the sample. In this technique the food sample and diluent are put into a sterile plastic bag which is then vigorously pounded on its outer surfaces by pads inside a machine called a 'Stomacher'. The resulting compression and sheering forces effectively breakup solid pieces of food. After samples are removed for analysis, the bag and its remaining contents can be discarded and the machine is immediately ready for reuse.

The stomacher is not as efficient as the blender is for breaking up fat globules, so that results on foods with fat levels above 20% are lower with the stomacher. Addition of 11% Tween 80 or other non-toxin surfactant will overcome the problem. Another factor highly important to accurate enumeration is nature of the diluent used. Mostly diluents used distilled water, saline solution, phosphate buffers, Ringer's solution, are toxin to certain micro organisms, particularly if the time of contact is unduly prolonged. A general purpose dilution medium in widespread use is 0.1% peptone. Physiological saline with 0.1% peptone added proved to be the most reliable. Peptone (0.1%) in saline (0.85% NaCl) is recommended by the international standards organization in its proposed procedure for aerobic counts at 300°C.

Inoculation of media should be carried out within 15-30 minutes of the preparation of the dilutions. It is preferable to prepare the initial 1:10 dilutions by using a joint glass long tube 100ml for 10g of sample or in 500ml for 50g sample. Subsequent dilutions can then be prepared volumetrically by pipette. Calculation of counts can correctly be expressed per gram of sample. The more usual method of preparing a 10^{-1} dilutions by adding 90ml of diluent to 10g of food introduces an uncertainty into calculation as the extent of solubility of the food and the food volume are unknown. It should be noted that the use of refrigerated diluent when the sample is not at refrigerator temperature may cause a drop in count as a result of cold-shock.

DILUENTS FOR ANAEROBES

When examining a sampling for anaerobic bacteria either qualitatively or quantitatively reinforced clostridia medium or a similar formulation must be used as the diluent in order to maintain a low oxidation-reduction (redox) potential. Dispersion of the sample in the diluent should be achieved by a method that provides least opportunity for introducing oxygen into the mixture. For example the cloworth stomacher is better than a top driven homogenizer unless the sample container is flushed with oxygen-free nitrogen before blending.

Whenever very oxygen sensitive anaerobes are to be examined, as well as appropriate diluents, special precautions need to be taken such as using the Hungate technique or use of an anaerobic workstation.

Diluents for osmophiles and halophiles

A suitable diluent for osmophilic counts is sterile 20% sucrose solution. For investigation of halophilic organisms (e.g. in samples of curing brines) Sterile 15% sodium chloride can be used as the diluent.

Diluents from liquid samples

Pipette aspectically 10ml. of thoroughly mixed sample into a sterile glass tube with ground glass stopper and make up to 100ml with sterile diluent to give a 1:10 dilution v/v. The other method is to weigh with aseptic precautions 10g of thoroughly mixed sample into the tube and make up to 100ml with sterile diluent to give a 1:10 dilution v/v. Prepare further decimal dilutions as necessary in the usual way. Inoculation of media should be carried out within 15-30min. of the preparation of dilutions.

PLATE COUNTS OF BACTERIA

Counts of viable bacteria are commonly based on the number of colonies that develop in nutrient agar plates which have been inoculated with known amounts of diluted food and then incubated under prescribed environmental conditions. Such counts are sometimes erroneously called total plate counts, when in fact only those bacteria which will grow under the particular environmental conditions selected can be counted.

Multiple tube technique may be used if low concentrations of bacteria are expected.

Some of the organisms may be capable of good growth at 37°C but not at 20°C, some may grow at 20°C but not at 37°C and other may grow at both temperatures.

Similar problem occur when attempting to choose media. Colony counts and multiple tube counts assess viability as the proportion of the population capable have continued multiplication (i.e. colonial viability) in the incubation conditions provided. In other words the development of a colony (on or in an agar medium) or of turbidity indicates colonial viability in the environmental conditions found either in the normal post-processing state of food or in the consumer.

This distinction is especially important when stressed or damaged organisms are encountered in for example, processed or preserved foods. Thus if a food sample be examined using both total (microscopic) count and viable count, a high total count in conjunction with a low viable count does not necessarily indicate that the majority of the organisms observed microscopically are dead, but perhaps merely that they are unable to multiply in the particular incubation environment.

In most test situations more than one incubation temperatures is likely to be employed:

- 0-10°C for counts of psychrotrops and psychrophiles
- 20-32°C for counts of saprophytic mesophiles
- 35-37°C (or 45°C) for counts of parasites and commonsales of homoiothermic animals

55-63°C or higher for counts of thermophiles, phychrotrophic and psychrophilic counts are best determined using surface count technique so that bacteria are not exposed to molten agar.

The degree of spoilage and probable shelf life of ice-packed chickens would best be best estimated by using an incubation temperature of 0-5°C, similarly the overall bacteriological quality of refrigerated farm bulk tank milk and / or the effectiveness and duration of its refrigeration would be best determined by a psychrotrophic count at 7°C for 10days, a count at 25-30°C would provide the most-information about general plant sanitation. Accurate control of the incubation temperature is essential in each procedure.

The thermoduric count measures the number of micro-organisms that survive a pasteurization procedure of 63°C for 30 minutes. Such test evaluates the build up of bacteria in blanchers or in other equipment held hot for long periods. Thermoduric organisms usually fail to grow at refrigeration or ambient temperatures.

DETECTION AND ENUMERATION OF INDICATOR AND INDEX ORGANISMS

The term "Indicator Organisms" are applied to those organisms who are contaminants of faecal (human or animal) or environmental origin and have reached food or water as a result of poor hygiene and sanitation. It is often associated with organisms of intestinal origin but other groups may act as indicators for other situations. For example the presence of members of the set 'all Gram-negative bacteria' in heat treated foodstuffs is indicator of inadequate heat treatment (relative to the initial numbers of these organisms) or of contamination subsequent to heating. Coliform counts, as the coliforms represent only a subset of "all Gram-negative bacteria" provide a much less sensitive indicator of problems associated with heat treatment, but are still frequently used in the examination of heat treated food stuffs.

The use of marker organisms in quality assessments originated in microbiological examinations of drinking water supplies in which intestinal commensals such as Eschrichia Coli were considered to be indicative of a potential hazard from the possible presence of intestinal pathogens.

For E.Coli to act satisfactory as an indicator (or index organisms as we call it today) in this situation.

(a) It must be present whenever the pathogens concerned are present.

(b) It must occur in greater number than the pathogens and

(c) It must not be less resistant than the pathogen to the envioirnment or to any process or manipulation imposed.

There are, however limitations to its use in foods where there appears to be little or no correlation between the presence of E.Coli and pathogens such as salmonella in meat, for example.

although E.Coli can not usually grow in water in temperate countries, it can grow in the richer environment provided by many foods.

The concept is based on Shardingen's suggestion in 1892 that members of the species we know call E.Coli be used as an index or indicator of fecal pollution since they can be recovered with less difficulty than Salmonella species. Other indicator groups or tests suggested or used include the fecal streptococci or enterococci, the Enterobacteriaceae staphylococci and the presence of Geotrichum, Candida, the machinery mould an indicator of plant sanitation and contaminated equipement. One criticism of using coliforms and faecal Coliforms is that their absence could give a false reassurance of safety when lactose-negative organisms predominate. The lactose organisms include, not only Salmonella and Shigella but also enteroinvasine strain of E.Col (EIEC) such as 014. For this reasons, tests for the whole of Enterobacteriaceae are increasingly being used. The Enterobacteriaceae includes even more genera of non-faecal origin than the Coliforms such as Erwinia and seiratra which are predominantly plant associated. For this reasons Enterobacteriaceae counts are used more generally as an indicator of hygienic quality rather than of faecal contamination.

It is concluded that inspite of the shortcomings, absence of E.Coli is still widely used as an indication of 'safety' of the food stuff.

TOTAL ENTEROBACTERIACEAE, COLIFORM ORGANISMS AND ESCHERICHIA COLI

Escherichia Coli (E.Coli) may be used with care as an indicator or index in food microbiology. Presence of E.Coli in water indicates faecal pollution of the water and may contain pathogens of intestinal origin. In addition serovars of E.Coli are enteropathognic. They are significant cause of infantile gastroenteritis, but may cause outbreaks of food poisoning amongst the adult population.

Coliform organisms including E.Coli some time present in raw food and uncooked ingredients and in due course these multiply. The presence of E.Coli and other enteric commensals in food stuffs may indicate the possible presence of enteric pathogens, the absence of the commensals cannot be taken as indicating the absence of such pathogens.

For example food material of animal origin (Meat, eggs, Milk) may contain salmonella in the absence of E.Coli if the animal had been suffering from the relevant salmonella infection. Therefore, one must be careful while interpretation of counts.

One problem with E.Coli is that it is less resistant than Salmonella in certain food containing nitrite or to certain processes. In such cases E.Coli derives from faecal contamination(handling of food by workers with dirty hands), only examination immediately after contamination will detect E.Coli in numbers proportional to the amount of faecal contamination.

In case sampling and examination is not carried out immediately after contamination, then depending on the nature of food and storage conditions, one of the following events may occur:-

(a) The E.Coli may die off so that the occurrence of the initial faecal contamination is undetected. In this case we may be incorrect in assuming that other entire pathogens will have died at the same rate.
(b) Total numbers of E.Coli remain the same. Again this cannot be taken to indicate that other entire organisms will react in the same way to the environments in which they find themselves.
(c) E.Coli grows and increase in number.. In this case we must assume that the conditions that were favourable for the growth and multiplication of E.Coli probably also favoured the growth and multiplication of other bacteria of enteric origin.

Thus although large number of E.Coli do not indicate recent or heavy faecal pollution, they must be taken to indicate a possible hazard from entire pathogens.

Nevertheless, the ability of E.Coli to survive and/or multiply cannot be used as a reliable indicator of the ability of pathogens such as salmonella to survive and/or multiply. It has been suggested by many Microbiologist that a better measure of the hygienic quality of foods would be a "Total Enterobacteriaceae count. Such counts are easily assessed by adding glucose to the selective coliform counting medium when violet red like agar is used, then glucose should be added up to 10% in the final concentration.

In this group of bacteria the following definitions have been assumed.

1. Total Enterobacteriaceae. Bacteria that, in the presence of bile salts will grow and produce acid from glucose (as determined by use of violet red bile glucose agar)
2. Coli-arogenes bacteria: Bacteria that, in the presence of bile salts or other equivalent selective agents, can grow and produce acid and gas from lactose when incubated at 30°c.
3. Coliform bacteria: Bacteria that, in the presence of bile salts or other equivalent agents, can grow and produce acid and gas from lactose when incubated at 35 to 37°c.
4. Faecal Coliform Bacteria: Bacteria that, in the presence of bile salts or other equivalent selective agents, can grow and produce acid and gas from lactose when incubated at 44-45.5°c. Note that the incubation temperature is critical and a digital incubator should be used for this purpose. The choice of temperature within the indicated range depends on the choice of medium. If equivalence of counts is used as the criterion, but it does not necessary follow that when the same counts are obtained by two different procedures, the same bacterial types are being detected.
5. Escherichia Coli: Bacteria that in addition to showing the above characteristics, are also methyl red positive, Voges, Proskauer negative, and cannot utilize citrate as a sole carbon source. Indole positive strains are termed E-Coli type I and are presumed to have the intestine as the primary natural habitat. In Coliscan MF method give blue-purple colour. E-Coli produces enzyme glucuronidase and not by Coliform.

The Verocytotoxigenic E.Coli 0157:H7 grows very poorly at 44°c, also many taxonomists consider Shigella to be taxonomically indistinguishable as a genus from E.scherichia, with some of the enteroinvasive serovars of E.Coli (EIEC) being equivalent to certain Shigella. The EIEC ferment lactose slowly or late or not at all, and are non motile(ICMSF 1996 report). Consequently

Verocytotoxigenic E.Coli (VTEC) and EIEC will not be detected by isolation techniques.

For total Enterobacteriaceal determination ICMSF1978 recommended that, to detect low concentrations liquid enrichment is carried out in phosphate-buffered peptone water or tryptone soya peptone broth followed by selective liquid enrichment in Enterobacteriace. Enrichment broth (containing oxgall and brilliant green) and plating on violet red bile glucose agar. Glucose colonies are checked to be oxidase negative. This method has been specified in 1SO8523:1991. For counting higher growth and omitting non-selective enrichment, similar procedure has been followed for both most probable number (MPN) and colony counts (1SO7402.1993)

In practice Total Enterobacteriaceae counts between 10-100 per gram or more are determined by preparing decimal dilution series and follow pour plate count method on violet red bile glucose agar (VRBGA).To each plate, 15ml of VRBGA is added, mixed well and allowed to set. The set medium is then overlaid with another 15ml of molten VRBGA at 45°c to supress the surface colonies. Incubate the plates after setting at 37°c for 24 hrs.

COUNTS OF COLIFORMS, FAECAL COLIFORMS AND E.COLI BY MULTIPLE TUBE TECHNIQUE

The production of gas and change in colour in the media described below indicates the presence of coliform organism. As many foods contain beside lactose other carbohydrate (e.g. sucrose, glucose, fructose) it is necessary that presumption positive results are confirmed by subculture. The incubation temperature may be used between 30° and 37°c. Obviously, the lower the incubation temperature in this range, the greater the count is likely to be.

MacConkey broth

Pipette 1ml of each of the prepared dilution of the food into each of three or five tubes of MacConkey's broth containing inverted Duhram tube. If very low concentration of organisms are expected, 10ml. or more amount of lowest dilution may be added.

ENUMERATION OF COLIFORMS

Incubate at 30°c, 35°c or 37°c and examine after 24 hrs and 48 hrs for growth accompanied by the production of gas and change in colour of the medium. The production of gas and acid (change in colour) after incubation is recorded as being presumptively positive. The production of gas (filling the concave of the Durham tube) without the need for testing for acid production usually regarded as indicating a presumptive positive result.

Presumptive positive tubes should be confirmed by subculturing two or three loopsful from each tube into a tube of brilliant green lactose bile broth or by streaking across MacConkey's agar plates. These are incubated at 35°c for 48h, gas production as described above being taken as indicating a confirmed result. By comparing the combination of positive and negative tubes obtained at each stage with probability tables(see appendix II) the most probable number of "presumptive Coliforms" and of "Confirmed Coliforms" can be determined.

Lauryl Sulphate tryptose broth at 30°c, 35°c & 37°c (ISO 4831:199.) An alternative procedure involves the use of Lauryl Sulphate tryptone broth with incubation at 30°c, 35°c or 37°c and examine after 24 and 48 hrs for growth and production of gas. The production acid may be detected by pH paper after incubation.

Enumeration of faecal Coliform and E.Coli by the tube technique

Facal Coliform Organisms that are able to produce gas and acid from lactose at 44-45°c (and which, for most purposes, can be regarded as equivalent to E.Coli) are confirmed by subculturing from presumptive positive tubes from the foregoing tests into EC medium incubated at 44.5 + 0.2°c for 24 h or Brilliant green lactose bile broth incubated at 44.0 + 0.1°C). for 24 hr. This is known as modified Eijkman test. It is advisable to prewarm the tubes to the incubation temperature before inoculation

Escherichia Coli 'type' is confirmed by indole test described under Biochemical tests. The confirmation of E.Coli by1MVIC tests (Indole production, methyl red tests, voges Proskauer test and

Citrate utilization) is time consuming because cultures for these tests are incubated at 37°c, it is first necessary to obtain pure cultures by streaking the appropriate broth cultures on to a solid medium(preferably one that is deferential but non selective). A number of enteopathogenic strains of E.Coli especially the E1EC and VTEC strains) will record to these tests negative results.

COLONY COUNTS OF COLIFORM BACTERIA AND E.COLI.

Plate count method

Medium used is violet red bile lactose agar (ISO4832-1991) or MaCconky agar. Duplicate sets of plates using 1-ml of the chosen range of dilutions of food. Add to each plate about 15 to 20ml of the media cooled to 45°c. Mix well and allow to set. Finally overlay with another 5ml of VRBA or MaCconkey agar. When the media has set, invert the plates and incubate at 35°c for 24 hr for coliform counts. Dark red colonies will appear. Generally typical coliform colonies will be 0.5 mm or more in diameter, overcrowding of the colonies will cause a reduction in size, confirm these colonies, by picking them off and inoculating lactose broth or Mac Conkey's broth or trypton broth and incubating at 35°c or 44-45°c for E.Coli as stated in Eijkman test described above. In trypton broth add 1 drop of Indole reagent. A bright fuchie red colour at the interface of the reagents is indicative of presence of E. Coli A yellow colour is a negative test.

Counts by use of dry rehydrable films.

A useful petrifilm (3M, St.Paul, Minnesota USA) product is available for counting E.Coli. The medium is based on violet red bile medium, but it also contains 5 bromo-4 chloro-3 indolyl-B-glucoronide to detect glucuronidase activity. Colonies, of E.Coli are blue with an adjacent bubble of gas. It is an AOAC (1995) method.

COUNTS FOR LIQUID SAMPLES BY MEMBRANE FILTRATION

Counts of Coliform and E.Coli and their differentiation by membrane filtration using Coliscan MF media.

Coliscan MF for the Identification and Differentiation of E.coli and General Coliforms From Liquid Samples

The patented Coliscan method is a means to separate and identify E.Coli (fecal Coliforms) and general Coliform members of the family. Enterobacteriaceae from other bacterial types.

Coliscan MF (membrane filter)medium is a nutrient liquid information which uses two color producing chemicals. One for the detection of the enzyme glucuronidase (produced by E Coli Strains but not by general Coliforms) and one for the detection of galactosidase. (produced by all coliforms, including E.Coli). A liquid sample containing E.Coli and other coliforms is passed through a membrane filter to leave individual bacterial cells on the membrane filter. The filter is then placed on a pad soaked with the Coliscan MF medium. When incubated at a suitable temperature, the cells will grow into colony forming units (CFU) on the surface of the filter. If Coliform (including E.Coli) colonies are present, they will produce the enzyme galactosidase which will react with its specific color-producing substrate in the medium and a water insoluble pink pigment will color the colony. If E.Coli colonies are present, they will also produce the enzyme glucuronidase which will react with its specific color-producing substrate in the medium and a water insoluble teal-green pigment will color the colony. However, since E.Coli produces both galactosidase and glucuronidase, those colonies will be a combination of the teal-green and pink pigments and will appear as some shade of blue-purple.

Therefore, the growth and identification of E.Coli and general coliforms is a simple one step procedure with the Coliscan MF method. Results can generally be read in 18-20 hours if incubation is in the range of 34-37°c and the overall accuracy and ease of use surpasses the widely used M-TEC method.

There is however always the chance that an unusual strain of bacteria may be present and result in false, negative or false positive readings. For example the E.Coli 0157:H7 does not produce the enzyme glucuronidase and so will appear as pink colonies indistinguishable from other general coliforms. And certain strains of other members of the family Enterobacteriaceae(some

Salmonella and some Shigella) may produce the enzyme glucuronidase, although they do not produce galactosidase and therefore appear as teal-green colonies. These teal-green colonies do appear different than the general blue-purple E.Coli, and fortunately another easy test screen for these various unusual bacterial types is available and may be used for verfication after the initial readings This is a spot test for the presence of indole, which is produced by virtually all strains of E.Coli, but very few general Coliforms or other members of the family Enterobacteriaceae. If the membrane filter containing the colonies that grew on the Coliscan MF medium is carefully lifted from the pad and placed on another pad which has been soaked with a reagent to detect the presence of indole, colonies that are indole positive will be indicted by the formation of a red (or magenta) zone around them within five minutes. Thus, all E.Coli (including the non-purple 0.157:H7) will have the red zone and non-E,Coli(such as the teal green Salmonella or Shigella) will not have the red zone. This simple additional test eliminates almost all possible false readings and adds further precision to the already excellent Coliscan MF method.

There are other tests available for either E.Coli or Coliforms which use single chromogens and which will identify either E.coli or coliforms, but not both in the manner which Coliscan does. The use of our patented process gives the best assurance that the colonies counted as E.coli are correct because Coliscan tests for the presence of the two enzymes produced by the great majority of E.coli, while single chromogen tests for E.coli detect only the single enzyme glucuronidase. Therefore, with Coliscan, only deep blue-purple colonies are counted as E. coli and the color is a combination of the two chromogens. General coliforms (which produce only galactosidase) are pink-red and glucuronidase only producers are a teal green color.

Sometimes waste waters or general surface waters also contain bacteria closely related to the coliforms which may grow on the Coliscan medium. Two that are common are Pseudomonas and Aeromonas species. Pseudomonas species are galactosidase and glucuronidase negative so colonies are not colored pink or purple, but they generally produce their own water soluble bluish, greenish or yellowish pigments which form a diffused colored zone around colonies. A heavy growth over a membrane filter may give the entire normally white filter a colored appearance. Colonies producing the soluble pigments described are very easily verified as non-coliforms with the oxidase test. A small drop of colorless reagent on the colony gives a dark purple zone around the Pseudomonas colony within 15 secods. Coliforms are negative and develop no color.

Aeromonas species are galactosidase positive and therefore grow as pink colonies like the general coliforms. However, if Aeromoas is suspected, they too can easily be differentiated from coliforms with the oxidase test. With the test reagent, Aeromonas colonies will develop a deep purple zone like Pseudomonas.

It is possible to eliminate these organisms from growing on the Coliscan medium through the use of various inhibitors and temperature control. However, they do not generally cause a problem with the general analysis. E.Coli will still grow as deep blue-purple colonies and debate continues on the importance of Aeromonas with many persons believing that it is as important an indicator as general coliforms. Likewise, the presence of Pseudomonas in water can be important from a sanitation and medical perspective.

- Coliscan has numerous advantages over the other methods for water testing:
- Incubation temperature is not critical. Accurate results occur over a range from ambient to 37°c.
- Coliscan visually differentiates and identifies E.coli and general coliforms from other members of the Enterbacteriaceae in one simple test procedure.
- Test results in 24 hours or less.
- May be used for other indicator organisms such as Pseudomonas and Aeromonas species without interfering with E.coli and general coliform growth and expression.

THE ENTEROCOCCI

Among the streptococci, Lancifield's group 'D' is composed to four species, namely, streptopcoccus faecalis, S.faccium, S.boris and S.enquinus. These are found in human beings and in cattle.

A fifth species, S. avium, possesses both the groups Q and D antigens and is physiologically similar to the group D streptococci.

Sherman (1937) classified the streptococci into four divisions according to their physiological and growth characteristics as well as tolerance tests. One of the four divisions he established was the enterococcus divisions, consisting of four species: S.faecalis, S.liquefaciens, S.zymogenes and S.durans. The name used to designate this division, as employed by Sherman, was intended to denote a specific taxon without regard to habitat or source. Sherman recognized that all members of his enterococcus division belonged to Lancefield's serological group D.

Subsequent to Sherman's classified review of the streptococci, numerous investigations contributed toward a cleared understanding of the species that comprise the enterococcci. Ultimately, the species were recorded into a more logical taxonomic scheme that comprised only two species and their respective varieties, namely, S.faecalis and S.faecium.

The revised system of specification in no way altered the taxonomic limits of the conterococci as designated by Sherman. The criteria for their identification as a specific taxonomic entity were retained.

These organisms (and especially E.faecalis and E.faeclium) are useful indicator and index organisms. Compared with E.coli they are more resistant to freezing, low pH and moderate heat treatment.

The enterococci may be enumerated in food as an aid to estimating the sanitary conditions under which it was produced.

Many foods are expected to contain enterococci part of their normal flora. Thus the enterococci may frequently be detected in frozen foods, fruit juices or foods that have a received a cursory heat treatment, even when E.Coli has been killed by the minimical conditions.

The enterococci are occasionally associated with disease, although generally they are not considered to be pathogenic. More appropriately they could be considered as "opportunists" and therefore are found associated with numerous low-grade infections.

These are possibly associated with food born-illness the entrococci may be the causative organisms in case of urinary infections, neonatel meningetis endocarditis, septicemia etc.

These organisms have been detected by plate counts using either Packer's crystal violet azide blood agar or maltose azide agar(KF streptococcus agar). These methods were described by the ICMSF (1978). The international commission on Food Microbiology and Hygiene (ICFMH) has proposed Kanamycin aesculin azide agar.

For smaller concentrations of organisms an MPN technique can be used by using maltose azide broth (KF streptococcus broth) or glucose number of enterococci and group D streptococci that are not necessarily of faecal origin.

The presence of azide in the media helps to inhibit staphylococci, as a result of its inhibitory action against the cytochromes.

Glucose Azide Broth

Procedure for small number of enterococci counting by Multiple tube technique.

Measure aseptically 1 ml of each of prepared dilutions of the sample and transfer to thrice or five tubes of glucose azide broth. If very low concentration of organisms are expected, 10-ml or even 100ml a mounts of the lowest dilutions may be added to equal volumes of double strength medium, thrice or five tubes being prepared at each dilution.

Incubate the inoculated glucose azide broths at 35°c for 72 h and examine for the production of acid. Record those tubes that are positive (i.e. in which acid is produced) and subculture a loopful from each positive tube into a fresh single glucose azide broth (5ml. per tube) and incubate at 45°c in a water bath for 48hrs or in digital incubator, examined the tubes after 18 and 48hrs. The production of acid at 45°c within 18hr indicates enterococci, and acid is produced after 18hr but within 48hrs, it is presumptive evidence of enterococci and can be confirmed rapidly by microscopic examination for the presence of short-chained streptococci. The most probable number of enterococci can be determined using probability tables (see appendix II).

Enumeration of Presumptive enterococci by using Kanamycin Azide (Kaa) Agar or Packer's Crystal Violet Azide Blood Agar or Maltose Azide Media

Procedure: Prepare food samples by one of methods described in preparation of samples (food homogenate). To duplicate sets of Petri dishes, pipette 1ml-aliquote of each of the decimal dilutions of the food-homogenate. Immediately add to each dish 15ml of Packer's crystal-violet azide blood agar. Or (KF streptococcus agar) or Maltose azide media melted and tempered to 44-46°c. Mix the samples and agar by rotating and tilting the dishes. After the agar has solidified, Incubate the plates inverted at 35-37°c for 72hrs (Packer's) or 48hrs (KF) or 48hrs Maltose Azide media.

On KAA agar small colonies of enterococci of white or grey and surrounded by large black haloes. Some strains of mesophilic lactobacilli may grow on this medium, and some of these are capable of splitting aesculin and therefore producing black holes. These can be distinguished by microscopic examination.

On Packer's agar small violet coloured colonies will grow. These should be subcultured by streaking across poured plates of Barne's thallium acetate tetrazolium glucose agar. Incubate at 35°c for 24hrs. Typical colony forms are: E.faecalis-colonies with a red centre and with or without a white periphery other group D streptócocci white or pink colonies. On maltose azide media dark red or have a red or pink central area will appear. Count the number of presumptive enterococci colonies and calculate per gram of the sample.

Confirmation of Enterococci

Select about 10 colonies from the packer's plates or an equal number of red /or light pink colonies from the KF plates. Pick individually into tubes of Brain Heart infusion broth and incubate at 35-37cfor 18-24hrs or until turbidity appears.

Prepare Gram's stain of each of the cultures and observe for typical Gram-positive, Oval Cocci, in pairs or short chains.

Remove aseptically about 3 ml. of each of the Streptococcus cultures and mix in another tube with about 0.5ml. of 3% hydrogen peroxide. Failure of bubbles to appear(catalase -negative)further confirms that the culture is a streptococcus.

Inoculate one tube each of Brain Heart infusion. Broth with the confirmed Streptococcus insolates. transfer the tubes in a 44-46°c water bath and incubate at 44-46°c up to 48hrs, and look for growth. Inoculate one tube each of Brain Heart infusion broth containing 6.5% Nacl with the confirmed Streptocoecus isolates. Incubate at 35-37°c for 72 hrs and look for growth.

A catalase-negative streptocoecus that grows at 44-46°c and in the presence of 6.5% Nacl confirms that the culture is an enter-occus.

Bacillus as indicator organisms

Two set of food, one have received pasteurization heat treatment and 2) have not permitted subsequent germination of surviving spores, post processing contamination may be detected by determining the proportion of viable bacteria present which are spore forming.

The procedure is to set up an aerobic mesophilic count on a food sample, and compare this with an aerobic mesophilic count after a laboratory pasteurization of a 10^{-1} dilution of the food.

Proecdure

1. Prepare a 10^{-1} dilution of the food in 0.10% peptone
2. Use this dilution to set up a decimal dilution series in a 0.1% peptone, and inoculate plates with 1ml. amounts. Prepare pour plates with plate count agar.
3. Immediately after preparing 10^{-1} dilutions, transfer 10ml. of this 10^{-1} dilution to each of two sterile test tubes. Insert a thermometer into one of the test tubes.
4. Place both tubes into a stirred water bath set at 81°c, so that the water in the water bath is a little above the level of the dilutions in the test tubes. Heat the tubes for 1min after the temperature indicated by the in-tube thermometer reaches 80°c. Remove the tube not fitted with the thermometer and cool rapidly under running water or in ice water.
5. Use the heated 10^{-1} dilution to prepare another decimal series in 0.1% Peptone, and inoculate sterile Petri dishes with 10ml

amounts. Prepare pour plates with plate count agar. This series of plates will provide an 'aerobic mesophilic spore count'.

6. Incubate both sets of plates at 30°c for 2 days, and then determine the aerobic mesophilic count and aerobic mesophilic spore count. If there has been no significant post processing contamination, the two colony counts should be the same.

The first category of foods for which this procedure can be used to determine the adequacy of the heat treatment and the absence of post processing contamination consists of foods in which surviving spores have not been able , or have not had the chance to germinate. This type of food include dried foods and food with a low aw and heat treated foods immediately subjected to freezing. at all stages of storage and distribution.

The method is applicable to on line in Milk processing plant. The laboratory examination would need to be performed without delay to avoid the chance for surviving spores to start to germinate.

Detection and Enumeration of Food born Diseases Bacteria

There are four main bacterial groups causing food-borne enteric infection in man. These are the salmonella, the shigella, the enteropathogenic, E.Coli and the Vibrio group. Staphylococcus may be an indicator of excessive human handling, a good poisoning hazard or a harmless contaminant of little significance.

There is possibility of food playing an important in spread of some viral diseases. Among viral diseases that may be food borne are infectious hepatitis A, poliomyelitis and enteritis caused by a range of viruses including rotavirus and Norwalk virus. To detect viruses in food the procedure is very expensive. A combination of immunocapture with polymerase chain reaction (PCR) amplification is likely eventually to provide an opportunity for the detection of small numbers of specific virus particles in sample of foods.

"Food poisoning" can be caused either by toxin produced by bacteria which may be non-viable when the food is consumed, or by the multiplication of bacteria in the intestine after consumption. When food poisoning is caused by ingestion of viable organisms (as in the case of Salmonella enteritis or clostridium perfringens food poisoning), the Laboratory procedure for assessing the hazard will depend on the use of highly selective media to detect the viable organisms. The hazard from toxigenic organisms e.g. staphylococcus aureus, Bacillus Cereus is based similarly on the use of selective media to detect the viable organisms. The toxin produced by these organisms is also a causative factor even though these are rendered ineffective by processing.

The presence of mycotoxins are estimated by bioassay techniques but chemical methods (by thin layer chromatography and high pressure liquid chromatography) are preferred.

In the case of pathogenic and toxigenic organisms it is necessary to use selective and differential media and the methods used can detect the organisms when they are in the presence of perhaps very large numbers of physiologically and / or ecologically similar organisms. The problem in detection is faced when the organisms have suffered some metabolic injury from food processing procedure as a result the organisms are inhibited by the very media that are expected to select for them. Yet because Salmonella in a frozen or dried food can not recover and grow in selective broth, for example it does not follow that the organisms could not recover in a non-selective environment. The selective procedures are relatively complicated and lengthy, liquid enrichment usually precedes the use of solid selective media, and not selective resuscitation may precedes the liquid enrichment stage.

When trying to determine the optimal isolation procedures it is available to use both resuscitation in order to ascertain which gives the best recovery rates from a particular food-processing storage systems.

Confirmatory and identificationn tests that follow the use of selective media are usually specially designed and offer confirmation only in the context of the prior selection of presumptive positive results. It is essential to ensure that the organisms submitted to confirmatory tests are obtained as pure cultures. For example, consider the case of a mixed microflora containing some coliforms organisms and some salmonella cultured on to desoxycholate citrate agar, a medium selective for salmonella which differentiates on the basis of lactose fermentation.

Coliform organisms grow only reluctantly on this medium and, if in associaton with a developing colony of Salmonella, they may not show their lactose-fermenting ability. The presumptive Salmonella colony transferred directly to confirmatory physiological test media is vary to show positive results characteristic of both the Salmonella and the coliform organisms. The purification of the Salmonella culture on a non-selective but differential medium would have demonstrated presence of Coliform organisms.

Following steps in isolation and identification are likely to be:

a. Pretreatment of a sample if necessary
b. Resuscitation in a 'non selective' non-differential medium
c. Enrichment in a liquid selective medium
d. Detection on solid selective and /or differential media
e. Purification of cultures on a differential nonselective medium
f. Confirmatory physiological tests and
g. Serological tests

Rapid and sensitive methods are being developed for salmonella and other pathogens in foods on the basis of nucleic acid based hybridization with a flow cytometry and cell sorting (FACS)

The convenient method of immunocapture involves the use of antibody immobilized on a surface are the dipstick. The dipstick is immersed in the enrichment culture and captured pathogens subsequently detected by an ELISA stage.

SALMONELLA AND SHIGELLA

Salmonella

All Salmonella are potential pathogens of a man. The examination of foods for their presence is very important. ISO6579: 1993 method of detection is described below:-

Six steps are followed :

1. Non-selective enrichment.
2. Selective enrichment
3. Plating on selective and differential media

4. Screening of suspect colonies on media revealing key biochemical characteristics of the organisms.
5. Antigenic analysis in two steps 1 a) the use of polyvalent O and Polyvalent H antisera. And (b) the use of group-specific O and H antisera.
6. Typing by means of bacteriophage.

Steps 1-3 for isolation and 4-6 for identification. Step 1 for the purpose of resuscitation (rejuvenating) Salmonella injured by processing or storage conditions. The second step is for selective enrichment, which favours the growth of Salmonella in an environment which may contain large number of bacteria other than Salmonella. Step 3 permits the 'sorting' of suspect Salmonella based upon biochemical characteristics. The fourth step permits further screening, based upon biochemical attributes. Step 5, through Seriological tests, leads to the ultimate identification of the suspect isolates as a member of the genus Salmonella. Step 6 method of refined identification of the isolate, such identification of particular value for epidemiological purposes, because serological types of a common bacteriophage pattern are likely to have a common phage, conversely isolate showing the same antigenic structure but dissimilar bacteriophage patterns are probably not of common origin.

The international commission on Microbiological specification of Foods (ICMF 1978) have recommended that all samples be pre-enriched in a non-selective medium prior to selective enrichment. They have recommended seven possible resuscitation media.

Isolation of Salmonella

1. Pre-enrichment of Salmonella. 25g of the food sample in 225ml of buffered peptone water (ISO 6579-1993) thoroughly mix , and incubate for 24 hrs at 37°c. This pre-enrichment medium is used for Dried whole eggs, dried egg yolks, dried egg whites, pasteurized liquid and frozen eggs,prepared powdered mixes, biscuits, infant foods, raw eggs, raw meats, eggs containing pasts, dyes and colouring substances, dried yeasts. If a dried food have a very high soluble solids contents (e.g. dried whole milk, dried baby food) prepare a non-selective resuscitation stage by adding

25g of food to sterile distilled water, making up the volume to 250ml. Incubate for 24hrs at 37°C. For candy and Candy Coating use 225ml. of reconstituted Nonfat dry milk with 1% Brilliant green 2ml, sterilize and adjust pH6.6-7.0

For animal by products and coconut use lactose broth with 1% Tergitol or Buffered Peptone water with 0.22%. Tergitol, 2.2.ml sterlize before use.

Note: If the sample contains a high concentration of sugar or salt, or other inhibitory substances, adjust appropriately the volume of the pre enrichment medium to promote the growth of Salmonellae. Similarly if the pH of the sample is so high or low that it may impair the growth of Salmonella in the pre-enrichment medium, adjust the pH with sterilized 1% suplhuric acid or 1% potassium hydroxide solution.

Salmonella Recommended Method

The pH of the inoculated medium should be 6.6 to 7.0 prior to incubation.

Selective Enrichment

Add 10ml of each pre-enrichment resuscitation culture to 100ml of selenite cystine broth, similar add 10ml of culture to 100ml. of Rappapory-Vassiliadis (RV) medium.

In case of foods not requiring a non-selective resuscitation stage add 25g of the food to 250ml of selenite cystine broth, similarly add 5g of the food to 50ml of RV medium and 25g to 250ml of Hajma's GN broth.Note that the use of the usual inoculum medium ratio of 1:10 impairs the efficiency of RV medium (Fricker, 1987)

Incubate the selenite cystine broth and Hajma's GN broth at 37°C for 48h and the RV medium at 42°C. When using 42 °C incubation it is important to prewarm the flask of pre enrichment and sample in a water bath before incubation.

Plating on selective Agar Media for Salmonella streak a 5mm loopful of from enrichment broth on the surface of one plate of each three selective agar media, brilliant green phenol red agar

(BGPRA), XLD Agar and Bismuth Sulphite agar or Hektoen Enteric agar.

Incubate BGPRA for 24 h at 35-37°C, XL D agar for 24 hrs at 35-37°C and BSA or Hektoen Enteric agar for 48h at 37°C. Examine for the presence of typical colonies, on BGPRA Colonies are Pink or red (Occasionally Colourless) surrounded by a zone of bright red medium. On XLD agar colonies are red with black centre. On BSA Colonies appear brown or grey to black. Sometimes with a metallic sheen. The medium surrounding the colony is usually brown at first, then turns black as the incubation period increases. Some strains produces green colonies with little or no darkening of the surrounding medium.

Identification of Salmonella

The suspected colonies require three stages to identify. Salmonella (I) Screening of suspect Colonies by use of determinative biochemical tests (2) Serological recognition of suspect isolates by use of polyvalent O, polyvalent H, group-specific O, and spicer-Edwards H antisera and (3) typing by means of bacteriophage.

BIOCHEMICAL TESTS

Pick off suspect Colonies at least five from each plate, and streak on to nutrient agar or a nonselective lactose agar to confirm the purity of the cultures. This step is most important to prevent physiological tests being performed on mixed cultures.

It is worth remarking that outbreaks of enteritis caused by lactose-fermenting variants of Salmonella have been reported (Poelma 1968, CDC1993) and that S.arizonae and some strains of Shigella will slowly attack lactose so that the purification stage should not be employed for rejecting cultures, especially if these have been derived as typical Colonies on isolation media not containing lactose e.g. Wilson and Blair's bismuth sulphite agar)

Pick off suspected colonies with sterile inoculating needle from each plate and inoculate the Triple sugar Iron Agar slants with 2.5-3 cm butts in tubes and lysine Iron Agar (LIA) slants with 2.5-3 cm slant. Inoculate the media by streaking back and forth on the slant and then by stabbing the butt.

Incubate the TSI and LIA slants 24h at 37°C. Salmonella suspect culture on TSI show alkaline (red) slants and acid (yellow) butts, with or without H2S production(blackening of the agar). Salmonella-suspect cultures on LIA show an alkaline (purple) reaction throughout the medium, with or without H2S production (blackening).

The suspect reaction on TS1 indicates that organisms ferments glucose (yellow butt) but fails to ferment lactose or sucrose (red slant). These reactions are typical of the salmonella. However, a few salmonella can ferment sucrose or lactose, to produce an acid (yellow) slant or butt.

The suspect reaction on 'LIA' in contrast results from the decarboxylation of lysine, producing an alkaline (purple) reactions in the butt of the tube. Organisms failing to decarboxylate lysine, but which ferment glucose, will produce an acid (yellow) reaction in the butt of the tube. Organisms which ferment neither glucose nor decarboxylate lysine will produce an alkaline (purple) slant and butt. Organisms showing this reaction are generally strict aerobes of the genus Pseudomonas. These isolates will fail to show growth in the butt of the tube, because of their aerobic metabolism.

Lactose and sucrose-fermenting Salmonellae will give typical 'suspect' reaction on LIA, since this medium contains neither lactose nor sucrose. Arizona organisms may give Salmonella-suspect reaction on TS1, due to late fermentation of lactose. These organisms will give Salmonella-suspect reactions on L1A, since they decarboxylate lysine and ferment glucose. More important, both Arizona and Salmonella which ferment lactose and/or sucrose will give typical Salmonella reactions on L1A, even though the same isolates may show non-salmonella-type TS1 slants. Similarly rare strains of Salmonella fail to decarboxylate lysine and this show an acid butt and an alkaline slant on L1A. These organisms show typical Salmonella reactions on TS1.

Other Biocheical Tests

Incubate at 37°C, 24-48 hrs.

(A) Urea agar-No change of colour of colonies.
(B) Tryptone broth-No indole.
(C) Dulcitol broth-Acid and gas.

SEROLOGICAL DIFFERENTIATION OF SALMONELLA (AOAC1963)

The presence of Salmonella in the sample culture may be confirmed by a slide agglutination test or Tube agglutination described in page no.85.

PHAGE TYPING OF SALMONELLAE:

Phage typing is a desirable epidemiological aid in the identification of Salmonellae Organisms for which acceptable. Phages are available. It is not possible for routine Laboratory to do phage typing because of costly antisera of limited shelf life, including single-factor antisera.

In these days various types of kits are available for identifications:

(a) Multi-coloured latex agglutination kits available from-spectate (Rhone-Ponlene) and wallcolex (welcome diagnostics) U.K.

In this kit a mixture of latex particles of three different colours, each colours being associated with particles on which are conjagated a different range of antibodies. The latex reagent is mixed with a bacterial suspension on a disposable 'slide'which provides a white ground. The colour of clumps of agglutinated particles against a particular uniform background colour of liquid (provided by unagglutinated particles) permits the identification of the Salmonella sero group. A dense suspension is to be used for colour development. The bacterial suspensions and latex reagents on slide is to be mixed thoroughly by rocking the slide manually or by electrical rotator.

The oxide Salmonella Rapid Test (SRT)

The oxide SRT is designed to combine liquid enrichment stage with liquid indicator or media, motile Salmonella being self-selecting by migration through the selective medium under the influence of chemoattractants. The test is performed in a disposable culture vessel with screw cap in which there are two tubes with porous basis. Each of the tubes contains two dehydrated media separated by a porous disc. The media in the bottom halves of the tubes are liquid selective media (RV medium) and lysine iron desoxycholate medium respectively) and the media in the upper halves of the tubes are indicator media (lysine iron cystine neutral red medium, and modified brilliant green lactose medium respectively) Presumptive identification of Salmonella is based on hydrogen sulphide production and lack of lactose fermentation.

To prepare the culture Vessel and tubes, firstly the dehydrated media are rehydrated. Then an elective medium and novobiocin are added to the culture vessel. The medium is inoculated with 1ml of a pre-enrichment culture of a sample. The Vessel is incubated at 41°C for 24h. The elective medium is designed to allow rapid growth of Salmonella whilst inhabiting growth of proteus, Edwardsiella and Gram -positive bacteria. During incubation the motile salmonella migrate into the tubes, and act on the media there. The colour changes occurring in the upper tubes are noted. From the upper layer a loopful of the presumptive salmonella culture is removed and tested against latex agglutination antisera.

The main advantages of the oxoid SRT is that only 24hrs of incubation is required after the pre enrichment culture and easy to perform.

Rapid Method for Detection of Salmonella by the Combination of Immunomagnetic separation and PCR (Polymerase Chain Reaction assay)

In comparison with conventional cultural methods, the IMS-PCR is a rapid and specific method. It takes 24h to detect Salmonella in non-fatty food samples. In PCR method a few copies of target DNA can be amplified to a level detectable by gel electrophoresis. But PCR can be inhibited by several factors (e.g. food components, humic acid, urine, bilesalts etc). The removal of inhabitory substances is a major step in the preparation of samples for PCR based detection of food pathogens. Immunomagnetic separation (IMS) is a powerful tool to extract bacteria from food samples. Bacteria are specifically separated from the specimen, resulting in a useful sample for PCR with little or no nonspecific DNA or interfering factors. During 1Ms, target bacteria from the pre-enriched sample are specifically caught on to magnetic beads coated with anti Salmonella antibodies. This complex of bacteria and beads is separated using a magnet and washed several times to remove food debris and other microorganisms, DNA of separated Salmonella is extracted and examined with real time PCR.

Polymerase chain reaction (PCR). Two pairs of oligonuclotide primers are prepared according to the sequences of chromosomal INVA and plasmid spvc genes. With these two PCR Primers, either one amplicon (from the INVA gene) or two amplicons (from the INVA and Spvc genes) are produced, depending on whether or not Salmonella contained a virulence plasmid. There are nearly 2200 Salmonella serovars and all of those tested so far seem to contain inva genes, which enable the bacteria to invade cells.

There are six salmonella serovars known to contain the virulence plasmid carrying spve C.genes namely: S.typhimurium, S.Choleraesuis, S.dublin, S.enteridis, S.gallinarum and S.pullorum. Except for gallinarum and S.Pullorum, which are specific for fowl, the other serovars named here are common etiologic agents of enteritis in humans.

The appearance of at least one fluorescent band or two bands by 1% agarose gel electrophoresis pre-stained with ethidium bromide, confirms presence of Salmonella.

This method is still on the stage of improvement especially for foods containing high fat contents that causes a loss in number of magnetic beads, which stuck to the food matrix and can not be separated by magnetic field.

Detection of E.Coli 0157: H7 (VTEC).

In many countries E.Coli 0157 has been found as a food pathogen. A number of serovars (e.g.026, 048, 0111, 0113, 0128, 0145, 0157, 0163) may produce verocytotoxins, but the serovars most commonly incriminated is 0157:H7. Since it produces serious nature of disease, specific screening for E.Coli 0157:H7 is important in food industry.

This organisms grows poorly at 44°C so the normal tests for Coliforms, E.Coli will fail to detect this organisms. This organisms is also normally sorbitol negative, where as most E.Coli are sorbitol positive.

PROCEDURE FOR DETECTION AND COUNTING BY PLATE MEDIA

The organism is grown on MacConkey's agar containing 1% Sorbitol (CT-SMAC) instead of lactose. The medium can be made more selective by adding cefixime (0.05 mg per litre) and potassium tellurite (2.5 mg per litre). Plating is done as described in Plate count method. Plates should be incubated at 37°C and examined for non-fermenting colonies. Hafnia and Escherichia heremanii give similar colonies on this media, so presumptive E.Coli 0157 need to be confirmed serologically.

Liquid Enrichment Media

The sample is prepared in buffered 1% peptone water containing cefixime (0.05mg/L) cefsulodin (10mg/L) and Vancomycin (8mg/L). This should be incubated at 37°C for 20-24 hours followed by subculture on CT-SMAC.

The latest method is immunomagnetic separation from liquid medium followed by plating the magnetic beads on CT-SMAC. This technique uses magnetic beads coated with antibodies specific for the target for the target organisms. Beads are available commercially for detection of E.Coli 0157(e.g. Dynabeads made by Dynal). After incubation of enrichment medium, the beads are mixed into the enrichment medium and then collected magnetically. The bacteria-carrying beads are then washed, re-suspended and plated on to CT-SMAC, which is then incubated as normal.

E.Coli 0157 can be confirmed by using latix agglutination kit available in the market.

Vibrio

Vibrio is a genus, Gram-negative, oxidase-positive motile rod straight or curved. Main difference between Enterobacteriaceae is their positive reaction to the oxidase test, although Vibrio metschnikovi is oxidase negative. The main species of vibrio involved in food borne infections are V. Cholerae, V. Parahaemolyticus and V. Vulnificus.

V.Cholerae, the causative organism of cholera is found in fresh water, bottled water and various types of foods especially grown under soil. V.Parahaemolyticus is a rapidly growing faculative anaerobe, capable of growth through a temperature range of 15 to 43°C, a pH range of 5 to 9 and a NaCl concentration of 0.5 to 8%. The organism is easily isolated from stools of patients, but from foods and environmental materials isolation is sometimes difficult because vibrios resembling. V parahaemolyticus are widely distributed. An outbreak of food poisoning was reported by many countries due to consumption of sea foods.

The incubation period varies from 6-20h with a mean of approximately 15h. The outset of symptoms may be sudden and include diarrhea, abdominal pain, vomiting, dehydration and fever. The symptoms have been likened both to those of cholera and of bacillary dysentery, since mucus and blood may be present in the fluid stool.

Freezing is said to affect vibrio adversely although it has been isolated from frozen foods. The total number of this micro-organisms required to cause illness is 106-109 per gram.

Method for detection and isolation of V.Para-haemolyticus (ISO 8914 -1990)

Method-A:

1. Prepare food homogenate with 25g food sample and 225ml of alkaline saline peptone water (ASPN) to give a 10^{-1} suspension. Similarly prepare 10^{-1} suspension in salt polymyxin B broth (SPB)
   Incubate both enrichment broths at 35 or 37°C for 8-24h.
2. After 8-24 h, inoculate a loopful of each enrichment broth (without first shaking the flasks) on to plates of TCBS agar (Thiosulphate citrate Bile salts Vibrios).
   Incubate the TCB at 35 or 37°C for 18-24h.
3. Examine plates for typical V.Parahaemolyticus colonies which are smooth and 2-4mm in diameter with green or blue centres, or colourless with a green centre (i.e. not sucrose fermenting). Sucrose fermenting organisms form yellow or yellow- brown colonies).

4. Pick off and subculture typical V.parahaemolyticus colonies, streaking on to plates of salt (3% W/V) nutrient agar. Incubate at 35 or 37°C for 16 to 24h. Confirm as described below.

Method for identification and isolation of pathogenic vibrio species

The media ASPW and TCBS will in principal permit the isolation of many vibrio species in addition to V.parahaemolyticus. However, cholerae will tend to be over grown in ASPW by the more vibrios. Roberts etal. (1955) suggested using a modified ASPW (MASPW) by reducing sodium chloride in 1% and adding also 0.4% magnesium chloride and 0.4% potasium chloride.

Method B

1. Prepare food homogenate by adding 25g of food sample to 225g of MASPW. Mix with a Stomacher. Incubate at 35°C or 37°C for 8-24h.
2. After 8-24h, inoculate a loopful (without shaking the flask) on to TCBS agar and on to modified cellobiose polymyxin B Colistin (McPc) agar.
   Incubate the TCBS at 35 or 37°C and MCPC at 40°C for 18-24h.
3. On TCBS agar, V.Cholerae, V.fluvialis. V.furnissi and V.metschnikovji are sucrose positive (yellow colonies) whereas V.vulnifus and V.parahaemolyticus are sucrose negative (blue-green or green colonies)

On MCPC agar, colonies of V.cholerae are purple whereas the cellobiose fermenting-vulnificus gives flat yellow colonies. Other species of vibrio grow poorly on MCPC agar.

Confirmation test of cultures from isolation methods 1 and 2:

1. Pick off and subculture selected colonies on to plates of salt (2.5% W/V) nutrient agar. Incubate at 35 or 37°C for 16-24h.
2. Identify purified cultures by
   (a) Oxidas test
   (b) B-galactose test

(c) motility test
(d) triple sugar iron (TS1)agar containing 2.5%(W/V) sodium chloride, streak the Butt and Slope.
(e) Tryptone water containing 2.5% (W/V) Sodium Chloride, for Indole test.

On TS1 agar, Vibrio produces no gas and no blackening (hydrogen sulphide production) V.parahaemolyticus. on TS1 agar produces a yellow (acid) Butt and red (neutral or alkaline) slant, Other Vibrio species also produce a yellow butt, but some will in addition produce a yellow (acid) slant because of the metabolism of sucrose.

Improved method for detection of Vibro parahaemolticus in sea food

In this method, sea food samples are cultured in salt Trypticase soy broth, which is non selective medium, and then a portion of the culture was cultured with salt polymyxin broth, which is a selective medium for V.paraharmolyticus.

This two-step enrichment was more effective than the one step enrichment in salt polymyxin broth alone. The enrichment cultures are then plated on to a new chromogenic agar containing substrates for beta-galactoseidase. The V. parahamolyticus Colonies developed a purple colour on this growth medium that distinguished them from other related bacterial strains.

It has been observed that V.parahaemolyticus colonies on TCBS agar are difficult to distinguish visually from other bacterial colonies, since they can be covered by a yellow colour produced by sucrose fermenting bacteria. To develop a more efficient method, we studied the effectiveness of a method involving two step enrichment with nonselective and selected media and plating on a chromogenic agár medium.

Procedure: 25g portion of seafood is taken, cultured with 225ml of tryptic soy broth supplemented with 2% NaCl (TTSB) at 37°C for 6 to 18h. One ml of the NTSB culture (Trypticsoy broth supplemented with 2% NaCl is transferred to 9ml of salt polymyxin broth (SPB) and incubated for 18h at 37°C.

A loopful of enrichment culture is streaked on to N-TCBS and CV (CHRO Magar Vibrio) agars. After incubation at 37°C for 18h colonies on C agar are violet colour shows the presence of V.Parahaemolyticus and green colonies on N-TCBS agar. The growth of colonies on CV agar is superior than on N-TCBS and stable in colour.

References: 1. National Institute of Infectious Diseases, Tokyo 162-8640, research paper.

2. Oliver, J-D and J-B Kapur 1997 Vibrio Species, Food microbiology: Fundamentals and front, American Society for Microbiology,Washington D.C.

6

Isolation and Enumeration and Identification

CLOSTRIDIUM PERFRINGENS

The presence of clostridium perfringens in foods produces toxin which causes food poisoning. It is designated as a heat resistance and produces spores which survive during cooking of foods. The main symptoms are abdominal pain, nausea and acute diarrhea. The effect of poison commence after 8-24 h after eating the contaminated food, mostly precooked meat or poultry. The spores can survive many hours of boiling and other method of cooking but can be destroyed by steam under pressure.

Isolation and Enumeration and Identification

Procedure for Colony Count

Pipette asesptically 1ml of each dilution (10^{-1} to 10^{-6}) of food homogenate prepared in reinforced clostridium medium or in Ringer solution of quarter strength as the diluent carryout plate counts on the dilutions using egg-yolk-free tryptose. Sulphite iron citrate cycloserine agar (S1CA) (1SO7937:1977), Shahidi and Ferguson's polymyxin Kanamycin Sulphite agar (SFA) (1CMSF, 1978) or neomycine blood agar. When neomycine blood agar is used the plates should be pre-poured and pre-dried and surface inoculated.

Incubate plates anaerobically in an aerobic jar at 35 or 37°C for 25h. On S1CA and SFA medias, colonies are coloured black (due to the reduction of sulphite causing a precipitation of Iron Sulphite). On neomycin blood agar (horse blood) colonies of Cl.perfringens produce a narrow surrounding narrow zone of complete haemolysis and a further surrounding narrow zone of incomplete haemolysis.

Select-typical, well-isolated colonies of presumptive Cl.perfringens and stab into tubes of nitrate peptone water plus 0.3% agar and also streak on plates of wills and Hobb's lactose egg-yolk milk agar for Nagler reaction and lactose fermentation. Mark each plate in half on underside. On one half spread 3 to 5 drops of Cl.perfringen diagnostic antitoxin and allow drying. Streak the organism under examination over the whole plate. Also streak a plate of blood agar.

Incubate the tubes of nitrate agar and plates of Willis and Hobb's medium anaerobically for 24h at 37°C. Incubate the blood agar plates aerobically for 24h at 37°C.

After incubation, examine the nitrate agar for the type of growth (indicating motility back of it) and test for nitrate reduction. Examine a Gram-stained smear prepared from the growth on Willis and Hobb's medium. Gram positive non-motile rods, capable of reducing nitrate are Cl.perfringens. This is confirmed by

(a) The reaction on Willis and Hobb's medium-Cl.perfringens gives cloudy zones around the colonies on the section of plate without antitoxin, but no zones on the section with antitoxin positive test devotes the production of a toxin (lecithinase C). The medium surrounding the Colonies will be red in colour as a result of lactose fermentation
(b) No growth on aerobically incubated blood agar plates.
(c) Perfringens by MPN method

Using food suspension prepare decimal dilutions (10^{-1}, 10^{-2}, 10^{-3}) using differential reinforced clostridia medium (DRCM) as selective medium and diluent. Pasteurize at 63±2°C for 15min. This will destroy vegetative cells, but will heat activate and encourage germination of spores within the incubation period.

Consequently spores and vegetative cells can be differentiated by the two MPN counts. Incubate at 35 or 37°C for 5 days. Tubes showing blackening should be confirmed for Cl.perfringens as described above method. Calculate MPN of Cl.perfringerns using probability table (Appendix II)

Rapid method of detection by PCR

A multiplex polymerase chain reaction (PCR) assay, developed to detect alpha-toxin and enterotoxin genes (cpa and cpe, respectively) of clostridium perfringens was used to identify enterotoxin isolates of this organisms from faeces and intestinal contents of pigs and from feed samples from pig forms.

The organism was grown on tryptose-sulfite-cycloserine (TSC) agar, TSC agar without egg-yolk, sheep blood agar, or in brain heart infusion broth or cooked meat medium. DNA was extracted by boiling and the PCR assay was carried out using reagents from a commercial kit available in the market. The 319 bp amplification product of Cpa and the 364 bp product of cpe were visualized under UV light after electrophoresis in a 2% agarose gel containing ethidium bromide. The average sensitivity of the assay, determined on artificially contaminated faeces was 9.2×10^4 (9.2×10^4) colony forming unit per gram.

PCR is a useful assay for rapid diction of Cl.perfringens in feed and for Confirmation of the identity of isolates presumed to be Cl.perfingerns.

Oxoid Ltd. method for detection of Cl.perfringens

This method is useful for detection of this organism in water. A new m-cp medium-a selective, chromogenic medium for rapid identification and enumeration of clostridium perfingens in water samples.

Compared to above stated methods, m-cp medium provides faster results with increased selectivity and specificity. Clostridrum preferingens spores are resistant to environmental stress and can survive in water for longer than vegetative bacteria, including E-Coli. This makes Cl. perfringens an important indicator of water pollution and a useful maker to alert the possible presence of other stores-resistant pathogens, such as viruses and protozoal cysts. In addition, its resistance to chlorination is useful in testing the effectiveness of water treatment processes.

The new oxide m-cp medium is designed to improve the differentiation of Cl.perfringens from other clostridial species and background flora. Chromogenic compounds within m-cp medium cause Cl.Perfingens colonies to turn yellow (based on their ability to ferment sucrose) thus differentiating them from other clostriduim species, whilst colonies of contaminating organisms turn purple (based on their ability unlike Cl.Perfringens to hydrolyse Indoxyl-b-D-glucoside). Additional conformation is provided by exposing the culture plate to ammonium hydroxide. This highly specific reaction causes acid phosphatase-producing Cl.Perfringens colonies to turn a distinctive dark pink.

The addition of D-cycloserine and polymixin B, and an incubation temperature of 44°C improve selectivity by inhibiting the growth of Gram-negative bacteria and staphylococci.

The chromgenic agar substrates in m-cp medium are so specific that further verification steps are not required. This allows results to be obtained in a day instead of 3-4 days required by traditional methods. The European council directive 98/83/EC recommends m-cp medium for testing water intended for human consumption.

COMPYLOBACTER JEJUNI AND C.COLI

Campylobactor Jejuni and C.Coli are Gram-negative, spiral-shaped, highly motile, microaerophilic organisms present in raw milk or inadequately pasteurized milk, river water and inadequately cooked poultry and pigs.

C.Jejuni and C.Coli cause gastroenteritis which is characterized by abdominal pain, fever and diarrhea. The severity may very from mild to dysentery-like symptoms with blood, mucus and leucocytes in the stools. The incubation period is 2-7 days and illness lasts 2-7 days. The organisms may be excreted in faeces for several weeks.

Because food may contain only a few cells, liquid enrichment methods are normally required before selective plating to detect contamination with C.Jejuni or C.Coli. A variety of enrichment broths and selective agars are prepared, a mixture of antibiotics and incubation under microaerophilic conditions at 42°C. Microaerobic gaseous atmosphere consists of 5-10% oxygen, 10% carbon dioxide and 85-90% nitrogen approximaely. This mixture is most easily obtained by using anaerobic jar without a catalyst, partially evacuating jar and refilling to atmospheric pressure with a nitrogen-carbondioxide mixture, using gas adsorbing/generating sachets designed for jar without catalyst (available from Oxoid). Dehydrated media and ready prepared antibiotic supplements are available from Oxiod and Difco laboratories.

Detection of C.Jejuni in foods that uses a short selective enrichment culture, a simple and rapid isolation procedure using nucleic acid sequence-based amplification (NASBA)

Abstract:-Interference by food components was eliminated by centrifugation following enrichment step. Subculture on NASBA media and incubate at 37 degree Celsius for 24hr. The detection of E.Jejuni is possible up to a ratio of indigenous flora to C. Jejuni of 10,000:1

It takes only 26h to complete the analysis

Ref:-Depth of food Technology and nutrition, faculty of Agriculture and applied Biological sciences university of Cohent, Belgium.

LISTERIA MONOCYTOGENES

Introduction:-Listeria monocytogenes is a motile non-spore-farming Gram-positive rod, generally found in soft cheeses, meat products, vegetables and seafood. Food-borne out breaks, do occur. Epidemic and sporadic outbreaks of listeriosis have been observed among domestic animals as sheep, cattle and poultry, less frequently among swine and dogs.

L. monocytogens can grow on cabbage and meat and in milk at temperature as low as 5°C and also survive in many foods for longer periods of time Regular pasteurization of milk will normally kill Listeria present in milk though those cells located inside polymorphonuclear leucocytes have a higher chance of survival at marginal pasteurization times and or temperatures.

METHODS OF DETECTION AND ENUMERATION

The method for detection of L. monocytogenes as per ISO 10560-1993 for milk & milk products and as general the method for detection (ISO11290-1-1997) and enumeration (ISO 11290-2-1997) :-

Plate method:

1. Weight out 25gm of food sample add 225ml of Fraser broth base to forms the 10^{-1} dilution and mix in a stomacher.
2. Prepare dilutions 10^{-1} to 10^{-5} in 0.1% peptone
3. Surface inoculate 0-1ml amount of dilutions on to prepared predried plates of polymyxin acriflarin- lithium- chloride- ceftazidihe- aesculin- manitol (PALCAM) agar and oxford agar. Retain the 10^{-1} dilution for the detection procedure.
4. Incubat the plates at 37°C for 48h. (the oxford agar plates aerobically) and the PALCAM agar plates under the micro aerobic conditions obtained as stated in method for enumeration of e.Jejuni.

DETECTION BY ENRICHMENT OF SMALL NUMBERS OF LISTERIA

1. Take 10^{-1} dilution from 225ml and add 2.25ml of sterile solution of lithium chloride, 0-23ml of a sterile of nalidixic acid sodium salt, 1.12ml of a sterile solution of acriflarvine, and 2.25ml of a sterile solution of ammonium iron citrate. Mix to form the primary enrichment in Half Fraser broth. Ready media also available. Incubate at 30°C for 24h.
2. Streak a loopful of the above broth on to a plate of oxford agar and an other loopful on to a plate of PALCAM agar. Incubate the plates at 35°C or 37°C for 48h (the oxford agar plates aerobically, and the PALCAM agar plates under the micro aerobic conditions obtained as explained in the enumeration of C.Jejuni.
3. Transfer 0.1ml of the primary enrichment broth to a tube containing 10ml of Fraser broth to give a secondary enrichment. Incubate at 35°C or 37°C for 48h.
4. Using the secondary enrichment medium and do the plating as per steps described in 2 above.

The Primary or Secondary enrichment culture can be screened in short time using an appropriate ELISA kit (e.g. TECRA Listeria Visual immunoassay).

IDENTIFICATION AND CONFIRMATION OF LISTERIA

1. Listeria hydrolyses aesculin, and all species of Listeria except L. grayi cannot produce acid from manitol, colonies of Listira on oxford agar are 2-3mm in diameter with dark brown or black haloes (due to aesculin hydrolysis); the colonies often have sunken centres. Colonies of Listeria on PALCAM agar are 1.5- 2mm in diameter, green and have black haloes.
2. To confirm Listeria, pick off five typical colonies and subculture each on to tryptone Soya yeast- extract agar (TSYEA), streaking to obtain separate colonies to ensure purity of culture, incubate aerobically at 35°c or 37°c for 24-48h.
3. Examine the colonies of each isolate that have developed on TSYEA, using a glowing mirror or Stereo microscope which appear blue, blue green or blue-grey with a finely granular surface resembling coarsely ground glass.
   Select an isolated colony on TSYEA and inoculate a tube of tryptone Soya yeast extract broth (TSYEB) and on the slope of TSYEA. Incubate at 25°c for 18-24h.
4. Gram staining and motility test to be performed on TSYEB. Listeria shows a tumbling motility. Listira is Gram positive, slim, short and non-sparing rods.
5. Perform a catalase text and oxidase test on growth from TSYEA slope. Listira is Catalase positive and oxidase negative.
6. The broth culture to be tested by inoculating with D-glucose, D-salicin, L-rhamnose and D-xylose fermentation broths medium containing tryptone 10%, sodium chloride 5% and Bromcresol purple 1% solution. Incubate at 35°c or 37°c for 7 days. Examine daily for acid production change from purple-yellow.
7. Nitrate peptone water for nitrate reduction, incubate at 35°c for 5days. Commercial kit AP_1 Listira kit (bio Merieux) may be used.

Results: Bacteria in the genus Listeria are Gram-positive, small slender rods, catalase positive, oxidase negative, produce acid from D-glucose, acid from D-salicin, and are motile with a tumbling motility.

STAPHYLOCOCCUS

Introduction:-The genus staphylococcus comprises three species, S. aureus, S.epidermidis and S. saprophyticus of which S. aureus is important in food microbiology. Staphylococcus food poisoning is an intoxication that depends on the ability of food concerned to support the growth of the staphylococci which produce the toxin. The toxin produced can withstand heating 100°c for 30 minutes and therefore absence of viable organisms in the food is not proof of safety if there is a high total (microscopic) count of cocci. The food preserved in high content of salt and also pickled food may allow growing S.aureus. The high content in processed food is a good indicator of the personal hygiene of factory workers, also due to respiratory infections of workers and other disease the food is infected. The symptom of food poisoning by staphylococci is nausea, vomiting, diarrhea, general malaise and weakness. The toxin produced by these organisms is called enterotoxines A, B, C, D and E.

Most food poisoning staphylococci are coagulase positive i.e. produce an enzyme which can clot rabbit or human blood plasma.

The Baired-Parker's medium and liquid enrichment in Giolitte and cantoni's tellurite mannitol glycine broth followed by steaking on milk salt agar (ICMF, 1978) are used for enumeration for S. Aureus. The method described in ISO/DIS6888-1 uses Baird-parker's medium followed by confirmation of production of coagulase and thermostable nuclease. The method in ISO/DIS 6888-2 examines for coagulase positive staphylococci directly on the isolation medium by using instead of Bair-Parker's medium, a rabbit plasma fibrinogen agar. This is stated to have the advantage of detecting the reportedly enterotoxigenic S.hyicus. This species produces atypical colonies on Baired-Parker's medium and does not produce a rapid reaction to the coagulase test, requiring the test to be incubated for 24h. It will therefore be missed by the method using Baird-Parkers medium followed by confirmation with the coagulase test.

MNP method is used for low concentrations of staph aureus in conjunction with Giolitte and Cantoni's medium.

METHOD (DIRECT PLATING)

1. Prepare dilutions (10^{-1}, 10^{-5}) of the food homogenate in peptone water.
2. Prepare plates of Baired-parker agar (20ml in each plate) and dry in laminar flow cabinet.
3. Pipette0.1ml of homogenate and dilutions of homogenates on to the surfaces of separate plates in duplicate and spread each portion with a sterile bent glass rod until the surface of the medium appears dry.
4. Incubate plates inverted at 35-37°c for 24h. Ifno colonies have developed, reincubate for a further 24h. Staph aureus typically forms colonies that are 1.0-1.5mm in diameter, dark black, shiny with narrow white margins and surrounded by clear zones extending into the opaque medium. There is a high probability that these colonies of staphylococcus aureus.
5. Pick off a number of colonies and test for coagulase production (see page 72). The absence of coagulase should not exclude food poisoning by Staphylococci as there have been reports of enterotoxin production by coagulase negative strains (omori and Kato 1959, Berg doll et al, 1971)
   MNP, Tellurite mannitol glycine Broth method

Method

1. Prepare samples as described in above method.
2. Pipette 1-ml portion from each of the first three decimal dilutions (10^{-1}, 10^{-2}, and 10^{-3}) into triplicate tubes of Tellurite Mannitol glycine broth. Carefully layer melted 2% agar to a depth of 2-3 cm on top of the broth to exclude air.
3. Incubate the tubes at 35-37°C for 24-48h, S.aureus will form a black precipitate or blacken the medium.
4. With sterile pipettes, remove drops of the culture from the bottom of each tube and spread them over the surfaces of individual Petri dishes containing milk salt agar. Incubate the plates at 35-37°C for 24hrs.
5. On Milk salt agar, colonies of coagulase positive staphylococci are smooth and round with entire margins which may or may not be surrounded by either an opaque or a clear zone. Select several suspect colonies from each and test them for coagulase production.

6. From the number of plates from each dilution containing coagulase- positive staphylococci, determine the MNP of S. aureus present in the sample by referring appendix II.

COAGULASE TEST (U.S FOOD & DRUG ADMINISTRATION METHOD JAN 2001)

Procedure

Transfer suspect S. aureus colonies into small tubes containing 0.2-0.3ml Brain heart infusion (BHI) broth and emulsify thoroughly. Inoculate agar slant of Triplicates soy agar (TSA) with loopful of BHI suspension. Incubate BHI culture suspension and slants 18-24h at 35°C. Retain slant cultures at room temperature for ancillary or report tests in case coagulase tests results are questionable. Add 0.5ml reconstituted coagulase rapid plasma with EDTA to the BHI culture and mix thoroughly. Incubate at 35°C and examine periodically over 6 h., period for clot formation. Only firm and complete clot that stays in place when tube is tilted or inverted is considered positive for S. aureus. Partial clotting must be tested further. Test known positive and negative culture simultaneously with suspect culture of unknown coagulase activity. Stain all cultures with Gram reagent and observe microscopically.

Ancillary tests

1. Catalase test use growth from TSA slant for catalase test on glass slide or spot plate, and illuminate properly to observe production of gas bubbles.
2. Anaerobic utilization of glucose. Inoculate tube of carbohydrate fermentation medium containing glucose (0.5%), Immediately inoculate each tube heavily with wire loop. Make certain inoculums reaches bottom of tube. Cover surface of agar with layer of sterile paraffin oil atleast 25mm thick. Incubate 5 days at 37°C. Acid is produced anaerobically if indicator changes to yellow throughout tube, indicating presence of S.aureus. Run controls simultaneously (positive and negative cultures and medium controls)
3. Anaerobic utilization of mannitol. Repeat 2, above using mannitol as carbohydrate in medium. S.aureus is usually positive but some strains are negative. Run controls simultaneously.

4. Lysostaphin sensitivity. Transfer isolated colony from agar plate with inoculating loop to 0.2ml phosphate-saline buffer, and emulsify. Transfer half of suspended cells to another tube (13 × 100mm) and mix with 0.1ml phosphate-saline buffer as control. Add 0.1ml lysostaphin (dissolved in 0.02m phosphate-saline buffer containing 1% Nacl) to original tube for concentration of 25μg lysostaphin per ml. Incubate both tubes at 35°C for not more than 2h, if turbidity clears in test mixture, test is considered positive. If clearing has not occurred in 2h, test is negative. S.aureus is generally positive.
5. Thermostable nuclease production: This test is claimed to be as specific as the coagulase test but less subjective, because it involves a color change from blue to bright pink. It is not a substitute for the coagulase test but rather is a supportive test, particularly for coagulase reactions. Prepare micro slides by spreading 3ml toluidine blue-deoxyribonucleic acid agar on the surface of each microscope slide. When agar has solidified, cut 2mm diameter wells (10-12 per slide) in agar and remove agar plug by aspiration, add about 0.01ml of heated sample (15min in boiling water both) of broth cultures used for coagulase test to well on prepared slide. Incubate slides in moist chamber 4h at 35°c. Development of bright pink halo extending at least 1mm from periphery of well indicates a positive reaction.

Some typical characteristics of 2 species of staphylococci and the micrococci, which may be helpful in their Identification, are shown in Table I.

Typical characteristics of S.aureus, S.epidermidis and micrococci (a)

Characteristic	*S.aureus*	*S.epidermidis*	*Micrococci*
Catalase activity	+	+	+
Coagulase production	+	-	-
Thermonuclease Production	+	-	-
Lysostaphin sensitivity	+	+	-
Anaerobic utilization Of glucose	+	+	-
Mannitol	+	-	-

\+ Most (90% or more) strains are positive,

\- Most (90% or more strains are negative

STAPHYLOCOCCAL ENTEROTOXIN

The ICMSF (1978) recommended two methods of extracting and concentrating enterotoxin in foods, followed by detection and identification of the enterotoxin by the micro slide gel diffusion test (also AOAC 1995, FDA 1992). The method is laborious, slow and inadequate in detecting low levels of contamination.

Rapid detection of enterotoxins in food using Transia Plate and Transsia Tube are two, state-of-the art "Sandwich" ELISA test kits are designed for the rapid detection of staphylococcal enterotoxins (SET) in food and bacterial cultures. Both test employ monoclonal antibodies that react specifically with food borne SET A, B, C, D, and E. The Transia kit is simple and very sensitive and results are available within there hours (sample preparation + ELISA). The kit, is ready to use reagents, which can be stored at least ten months, ensure flexible, swift, economical use.

BACILLUS CEREUS AND OTHER BACILLUS Spp.

Bacillus Cereus is a Gram-positive, facultative aerobic spore former where cells are large rods and whose spores do not swell the sporangium. These and other characteristics, including biochemical features, are used to differentiate and confirm the presence B. Cereus, although these characteristics are shared with B. Cereus Var. mycoides, B. thuringiensis and B. anthracis. Differentiation of these organisms depends upon determination of motility (most B. cereus are motile), presence of toxin crystals (B. thuringiensis), hemolytic activity (B. cereus and others are beta hemolytic whereas B. anthraces is usually nonhemolytic), and rhizoid growth which is characteristic of B. cereus var. mycoides.

A wide variety of foods including meats, milk, vegetables, fish and cereal foods have been associated with the diarrheal type food poisoning. Food mixtures such as sauces, puddings, soups, casseroles, pastries, and salads have frequently been incriminated in food poisoning outbreaks.

Method

A selective and differential medium such as mannitol egg yolk phenol red polymxin agar (MEPPA) is used in plate count method.

It is available in dehydrated form and is specified by ISO 7932:1993. However, MEPPA is readily overgrown and has poor buffering power against acid produced by mannitol fermenters (B. cereus does not ferment mannitol). It is suggested that parallel counts be set up on polymyxin pyruvate egg-yolk mannitol bromthymol blue agar (PPEMBA). Lecithinase activity and spore production are reported to be enhanced on the medium by the inclusion of the pyruvate and peptone, this medium is also have poor buffering power, so plates should be examined after 24h. of incubation. Presumptive B. cereus may be confirmed by Hugh and Leif son's medium to test for anaerobic dissimilation of glucose, hydrolysis of gelatin, reduction of nitrate, ability to grow on Kinsley's chloral hydrate agar, inability to attack mannitol or xylose.

Food can be examined directly for the presence of B. cereus enterotoxin. ELISA kit is available from TECA and a reversed passive latex agglutination kit is available from oxide.

Procedure

Prepare food samples and dilutions as described in the method of preparation and dilutions of food Homogenate using 1.5% peptone water.

Transfer 0.1ml from each dilution to duplicate plates of poured, well dried MEPPA and PPEMBA, spread the inoculums evenly over the surface of each plate with a sterile, bent glass rod, starting with plates at the highest dilution first. Incubate for 24h and 48h at 30°C.

On MEPPA presumptive B. cereus colonies are rough and dry with a violet-red background and a halo of dense white precipitate, on PPEMBA the colonies will be turquoise to peacock blue and surrounded by a precipitate. These reactions result due to B. cereus not fermenting mannitol and presence of lecithinase activity against the egg yolk.

In contrast colonies of B. subtilis and B. lickeniformis will be yellow and/or surrounded by a yellow zone (because of acid produced from mannitol) and show no zone of precipitation (because of lack of a lecithinase).

Confirmation Tests

To confirm colonies of B. cereus purity by streaking on egg-yolk agar plates, and prepare pure cultures on nutrient agar slopes. Examine microscopically for the presence of Gram positive rods and spores being ellipsoidal.

Biochemical tests

See page from 57 to 69.

(a) No acid is produced from mannitol
(b) Hydrolysis of gelatin (see page 57)
(c) Reduction of nitrate
(d) Ability to grow on Knisely's chloral hydrate agar.

Other tests are for strains of B. cereus are hydrolysis of starch and Voges-Proskaur positive.

Rapid identification kit VITEK or API made by Biomerieux France API 50 CHB is available in the market which takes 18-24h for identification of B. cereus and its isolates.

CLOSTRIDIUM BOTULINUM

Clostridium botulinum: Clostridium botulinum produces toxin which is highly potent and 2kg of this toxin can kill whole world. Food-borne botulism is a toxication due to consumption of food containing botulinum toxin formed during growth of C. botulinum.

Most outbreaks of botulism have been caused by smoked, cured, pickled or canned food. The organism generally is not enumerated in foods but efforts are mode on detecting the presence of one of the specific types of botulinal toxins in the suspected food, the laboratory undertaking this method must be secured and experienced Microbiologist handle this task.

The spores of Cl. Botulinum are relatively heat resistant but more sensitive to heat at low pH. As canned foods present suitably anaerobic conditions for the growth of cl. Botulinum if the pH is above 4.5, heat processing is designed to destroy the Cl. Botulinum spores.

Clostridium botulinum is an anaerobic Gram-positive rod whose spores are highly heat-resistant. There are seven known types (A to G) based on toxin production. During growth in food the organism produces a protein with a characteristic neurotoxicity, which is responsible for the intoxication. The seven types fall into three physiological and cultural groups, each having similar patterns of sugar fermentation, where as the remaining type Band F strains and all type strains are non-proteolysis. They ferment some of the same sugars as the proteolytic strains, but also a number that are different. The third group consists of types C and D, which are also non-proteolytic but ferment very few sugars. Types A, B, E and toxins are established as dangerous when ingested by man, but most outbreaks of human botulism have been caused by types A, B and E.

Types C and D have caused heavy losses to chickens and other fowl, cattle, horses and mink, but repots of human botulism caused by type C and D are rare (author and Lynt, 1972, Roberts et al., 1973).

SCREENING TEST FOR DETECTION OF CLOSTRIDIUM BOTULINUM

Toxin Identification by the screening procedure

1. Blend 10g of food stuff with equal weight of Gelatin phosphate diluents and homogenize in a mechanical blender. If the food sample is particularly solid or sticky, it may be necessary to homogenize, 1 part in 2 parts or more of the diluents.
2. Centrifuge the homogenate for 30 minutes in the cold.
3. Subject the supernatant extract to the following toxicity tests by intra peritonal inoculation of mice.
   (a) Pipette 1-2 ml in supernatant liquid into a small test tube and place the tube in boiling water bath for 10 minutes (caution, do not pipette by mouth) Inject one mouse with 0.5ml of the boiled supernatant liquid. This serves as a control test for heat-labile toxin.
   (b) Taking into account the initial dilution, make serial decimal dilutions of the food extract in Gelatin-Phosphate Diluent, upto 1:1000 if type E is suspected (e.g. fish, marine mamal)

and upto 1:100, 000 if type A or B is suspected (e.g. canned vegetables or fruits). Inject 0.5ml amounts of each dilution intra peritoneally into separate mice, and estimate the approximate number of mouse Minimal Lethal Doses (MLD) per ml or per Gram of the original food sample. Where the total Volume of supernatant extract is small, the volume injected may be as little as 0.2ml, but they should be kept constant in any given series observe test mice for up to 4 days.

(c) For more accurate estimates of toxicity, use a tripling series of dilutions e.g. 1:10, 1:30, 1:100, 1:1000 and 3-5 mice for each dilution. If no end point is thus reached, extend the series of dilutions and repeat. A decimal dilution series generally provides an adequate estimate.

4. Use additional mice for antitoxin-neutralization or type identification tests:-
   (a) Within 24h of performing the toxicity tests, it should be possible to estimate the dose of supernatant extract that contains about 10-30 mouse MLDS in 0.5ml (or in 0.2ml, if preferred). Use this amount as the challenge dose throughout, if the supernatant toxicity is too low to yield this amount in 0.5ml, challenge doses down to 3-10 MLD are permissible but neutralization test are unreliable at toxicity levels below this.
   (b) Inoculate each of two mice with the challenge dose of toxin. These controls should die with signs of botulism.
   (c) Calculate the amount (in suitable volume and dilution) of each monovalent antitoxin on hand that provides about 100 mouse-protective doses against homologous botulinal toxin. Inject the respective amounts of each antitoxin intraperitoneally into one or two mice. About 1h later, injects each mouse by the same route with the challenge dose of toxin. Alternatively, inoculate the antitoxin and toxin together, after mixing them in a small glass tube or directly in a syringe and incubating at 35-37°C for 30minutes.
5. Observe test animals for up to 4days. The presence of botulinal toxin is indicated by the occurrence, in mice unprotected by a specific antitoxin, of a progressive motor paralysis. When

large amounts of toxin are present the first signs may be apparent within 4-5h in the form of an indrawn or scapoid abdomen with "bellows" respiration, followed by dragging of hind legs and increasing weakness. With type E toxin, the period before the onset of symptoms tends to be shorter, and death generally occurs within 24h. The incubation period with other types of botulinal toxins may be prolonged to 8-10h and death may be delayed for a few days. Animals receiving toxin neutralized by botulinal antitoxin of a homologous type will survive, thus permitting the toxin type to be identified.

6. In routine examination of foods for toxigenic spores of C. botulinum the following preliminary procedure may be introduced before step above.

Incubate 5-10g of food material considered suitable for outgrowth of spores (e.g. fish, fish products, canned food, cooked poultry and meat products.) in test tube or glass sample Jar at 28-31°C for 7-10 days. Then proceed with steps 1-5.

Direct cultural procedure for Detection of clostridium Botulinum in suspect Foods.

The media used are (a) Cooked Meat Medium (b) Trypticase Peptone Glucose yeast extract broth (TPGY). For better results add trypsin to this medium (TPGYT). Toxin-containing cultures (pure or otherwise) should not be stored for future use on media containing trypsin, as continued action of trypsin may destroy the toxin).

Procedure

1. Take about 5g (5ml) of the food homogenate into each of three tubes of freshly boiled and cooked medium (Cooked Meat Medium or Tripticase Glucose Medium or TPGYT Medium if type E is suspected). Optionally, use all three Media.
2. If cooked Meat Medium or Trypticase Glucose. Medium is used, heat one of the tubes to 60°C for 15 minutes and another to 80°C for 30 minutes in water baths, Immerse the tubes to a depth greater than the level of material in them. Agitate tubes during the first 2 minutes of heating, leave the third tube unheated. If TPGY medium is used, do not heat the tubes, to do so will inactivate the trypsin.

3. Incubate the tubes at 29-31°C until vigorous growth is visible as evidensed by gas production, turbidity and in some instances partial hydrolysis of the Meat particles. As a rule toxin is present in the highest concentration after the period of active growth and gas production, turbidity, and in some instances partial hydrolysis of the meat particles. As a rule toxin is present in the highest concentration after the period of active growth and gas production, but may be detectable after 72h incubation. Generally not more than 5 days of incubation are necessary for optimal toxin production. To cover the possibility of delayed germination of C. botulinum spores, non-toxic culture tubes should be held in the incubator and re-examined after 10-14days. In then re-testing, to conserve mice a preliminary toxicity test is recommended, using 0.5ml of a 1:10 dilution of the supernatant portion of the medium, before proceeding with toxicity estimation or antitoxin-neutralization tests. Botulinal toxins may degenerate during prolonged incubation in trypsin-containing culture medium.
4. Test the supernatant liquid or sterile culture filtrate (using sintered glass or membrane filter) for botulinal toxins, as described in the above Screening Test for Detection of C. botulinum Toxin, steps 3.
5. The development of typical botulinic signs in unprotected mice, with protection conferred by type-specific antitoxin, confirms the presence of C. botulinum.

ISOLATION OF CAUSATIVE ORGANISMS IN IMPLICATED FOODS

Since the heat-liability of type E spores prevents the use of heat treatment to overcome gross contamination with vegetable forms of other species, the preliminary incubation may be followed by a 1 hour treatment of cultures with an equal volume of filter-sterilized absolute alcohol to kill vegetative cells with subsequent streaking on to Blood Agar and Brain-Heart infusion Agar.

PROCEDURE

1. Streak a loopful from each of the three cultures described above under Direct cultural procedure for Detection of C.

botulinum in suspect Foods (or from enrichment cultures prepared as prepared above) on to the surface of separate plates of Blood Agar and Brain. Heart infusion Agar or on either liver Veal: Egg-Yolk Agar or Anaerobic Egg-Yolk Agar or Reinforced clostridial Agar in such a manner as to obtain separated colonies. To prevent spreading growth as such as possible, dry the surfaces of the agar plates thoroughly before use.

2. Incubate plates anaerobically at 29-31°C. Colonies appear within l-2days and generally should be picked after 20-24 hr. of incubation. For characteristic appearances of C. botulinum colonies on Brain Heart and Blood Agar media, on the egg-yolk agar media colonies of C. botulinum exhibit a surface irridescence when examined by oblique light. The lustrous appearance resembles a 'pearly layer' and usually extends beyond and follows the irregular contour of the colony. In addition, colonies of types C,D and E ordinarily are surrounded by a Wide Zone (2-4mm) of yellow precipitate, colonies of types A and B generally show a smaller zone of precipitation. Difficulties in picking appropriate colonies may be expected since certain other clostridia form Colonies which exhibit these characteristics.
3. Select isolated colonies from the plates and inoculate cells from each into tubes of freshly boiled and cooled cooked. Meat Medium and Trypticase Peptone Glucose yeast-Extract Medium with added Trypsin.
4. Incubate at 29-31°C for 4-5days and confirm the type of organism serologically as described above, under screening Test for detection of C. botulinum Toxin, step 3-5. Page 173 to 175.

OTHER PATHOGENIC AND TOXIGENIC BACTERIA:

Aeromonas

Members of the genus Aeromonas are Gram-negative straight rod-shaped cells with rounded ends and coccoid forms. They are motile by polar flagella and facultatively anaerobic. There are three speices, A. hydrophila, A Sobria and A Caviae.. These may be found in untreated^ Water, raw shellfish and on foods washed in untreated water. Acremonas species produced toxin which showed partial immunological identity with Chlora toxin in the immunodiffusion procedure. The pathogenesis has been likened to that of vibrio parahaemolyticus.

Aeromonas should not be disregarded as a potential cause of gastroenteritis when isolated directly from food and faeces. The mechanism of illness may be similar to that of cholera.

In an acute case of gastroenteritis with profuse diarrhoea an almost pure culture of A. hydrophila was obtained on MacConkey agar. The colony was flat and resembled that of V. Cholera, biochemically the organism resembled a Vibrio.

Aeromonas can be distinguished from Vibrio in that it is resistant to Vibriostatic reagent 0/129(2,4-diamino-6, 7-di-isopropylpteridine) and does not require added sodium chloride in the medium.

Differentiation of three species on the basis of biochemical tests is shown in Table 1, in addition, the three show different reactions on blood agar.

Aeromonas hydrophila and A. sobria typically are both hemolytic on enquine blood agar, producing wide Zones of complete haemolysis. In case of A. hydrophila, however, the haemolytic zone is surrounded by a zone (halo) of precipitation, also in areas of heavy confluent growth of A. hydrophila the medium typically shows a green coloration.

DIFFERENTIATION OF THREE FOOD BORN

Human Pathogenic Species of Aeromonas Table 1.

Aesculin hydrolysis		*Gas from glucose*	*Acid from Salicin*	*V P. reaction*
A Hydrophila	+	+	+	+
A Sobria	-	+	-	d
A Caviac	+	-	+	-

Isolated colonies of caviae are non-haemolytic but show zone of precipitation. Haemolysis may be seen in areas of heavy confluent growth.

OTHER PATHOGENIC AND TOXIGENCE BACTERIA

Pseudomonas

Pseudomonads are ubiquitous in soil, on animals and plants and in water. Only a few species are hazardous for animals and human beings, whereas many are plant pathogens

Only two species, Ps. aeruginosa and Ps. Cocovenans are main food poisoning bacteria. The illness in adults is mostly mild enteritis but in infants it might lead to death after profuse diarrhoea.

A number of toxins (toxoflavin and bougkreak) is also produced by Ps. Cocovenans. The organism was isolated from a defective fermentation of bougkreak, a well-known food in southeast Asia. The bougkreak toxin is an unsaturated C29–fattyacid which inhabits oxidative phosphorylation and causes hypoglycemia and death in a few hours, or at least in a day or two (Van Veen 1966) Ps aeruginosa has created a serious problem in hospital, especially for patients with impaired defense against infectious diseases. Several hospitals survey's have revealed that vegetables, meats and frozen foods are sometimes contaminated with Ps. aeruginosa Human milk has also been reported as the source of infection of newborn infants. Ps is an organism of concern in drinking water especially in bottled natural mineral waters. The Pseudomonas have a respiratory, not fermentative metabolism. Some Species form pigments (green, blue, green, red, yellow, orange or light or dark brown) Their fluorescent pigments pyocyanin and fluorescein can be observed in spoiled foods by using an ultraviolet light.

Procedure

1. Prepare a decimal dilution series of the sample in 0.1% Pepton.
2. Prepare plates of media Cephaloridine-fucidin agar (CFCA) and inoculate on the surface of plates with 0.1ml amounts of dilution. Spread with a sterile glass spreader, starting with the

highest dilution. Incubate at 25°C for 48h. An incubation temperture of 41 ± 1°C can be used if seeking *Ps aeruginosa* preceded by a resuscitation stage at a temperature of 30°C for 4h, after which the plates are transferred to 41 ± 1°C.

CONFIRMATION

Colonies producing yellow green fluorescent diffusible pigment can be assumed to be Pseudomonas. Pseudomonas species not capable of producing fluorescent pigment will produce smooth translucent white or Cream colonies. Pseudomonas is (a) Oxidase positive (b) Catalase positive, (c) Oxidatively metabolizes glucose (determined with Hugh and Leifson's medium) (d) motile with polar flagella.

FOOD-BORNE PARASITIC PROTOZOA AND HELMINTHS

Parasitic Protozoa and Helminth (Soil born) are worldwide known for problem of food hygiene. These parasites come from the soil and animal feaces. The foods that are consumed raw or without cooking for example salad vegetables and fruits are the Vectors of these parasites. These remain in the crevices of food even after washing. They exist in fresh water or the oral cavity or intestines as animal parasities. The diseases they produce are varied and often have a worldwide distribution.

PROTOZOA

Among the protozoa, the most important are Entamoeba histolytica, Toxoplasma gondie, and Giardia lamblia.

The amoebae are microorganisms and can be detected in the laboratory. Some large parasites are easily visible to the naked eye, are transmitted by eggs which are microscopic. There are few accepted standard methods for determining precisely the numbers of parasites in a given samples of foods. No selective culture media for parasites from food exists.

Parasites are heat sensitive. Also most of the parasitic are sensitive to freezing, being destroyed by freezing to −18°C for 1 week. Laboratory examinations are useful to restrict outbreak other pathogenic and Toxigenic Bacteria of disease e.g. the nematode (round worms)) Anisaki simplex found in fish, can be observed naked eye using a good source of light passing through the food. However some times these parasites can easily be missed. Examination of stools samples under low power magnification worms can be seen.

The pork and beef tapeworms are human parasites which attach to the intestinal wall by means of such as and hooks on their rounded head, the scolen. The worms can grow up to 5 meters long. Infection is via ingestion of under cooked meat containing the larvae or by ingesting faeces containing eggs. Symptom include weight loss or inability to gain weight. Diagnosis is by detection of sectioned worms in stool samples. Flukes (Trematodes) causes oriental liver fluke infection. The organism is contracted by eating under cooked fish. Symptoms include stomach aches loss of appetite, enlargement of the liver and high blood count. Small, brown, ovoid eggs with a flattened and internal cilia may be evident upon microscopic examination of faecal samples.

CONFIRMATION

Colonies producing yellow green fluorescent diffusible pigment can be assumed to be Pseudomonas. Pseudomonas species not capable of producing fluorescent pigment will produce smooth translucent white or Cream colonies. Pseudomonas is (a) Oxidase positive (b) Catalase positive, (c) Oxidatively metabolizes glucose (determined with Hugh and Leifson's medium) (d) motile with polar flagella.

BRUCELLOSIS

Brucellosis is spread every where throughout the world. The people who come into contact with animals are especially effected. Veterinarians farmers, workers in slaughter houses, meat packing and processing plants as well as laboratory persons.

Infection in human beings occurs by ingestion of food, by accidental inoculation by inhalation or, most important by direct contact with animals, raw animal products, milk and milk products. Brucellace can survive in cheese, milk, in salted and otherwise cured meat for about six months. Therefore adequate heat treatment is the most reliable method of destroying brucellae in foods.

Brucellae are highly infectious, even in very small numbers, there is a high risk of laboratory acquired infections for all laboratory staff working with Brucella. Such work should be done only in specially constructed enclosed systems, that control air flow and incinerate the air before exhausting it.

For identification of Brucella in Milk, samples that possibly originated from infected animals Ring Test safely can be done as explained in serological tests page 86.It is advisable that Food Microbiologists working in food Laboratories should never attempt the isolation of Brucella from samples of raw milk or dairy products. Such investigations should be assigned to public health and clinical laboratories.

MYCOBACTERIUM BOVIS

Milk-borne tuberculosis in human beings generally is caused by Mycobacterium Bovis. Tubercle bacilli may be excreted with milk from Cows suffering from tuberculous mastitis. Yet. They may be present in milk even in absence of any visible pathological lesions of the udder. Pasteurisation of milk will destroy all strains of M. Bovis and M. tuberculosis. Wild animals can act as natural reservoirs of infection, therefore it is difficult to eradicate completely.

Mycobacteria is rarely cultured in routine by food. Microbiologist and it is advisable to do the investigation in public health and clinical pathology laboratories.

SHIGELLAE

Several plating media have been used for isolating shigellae. Each has advantages and limitations. Desoxycholate citrate, Hektoen Enteric and Salmonella-Shigella (SS) agars are highly selective and may restrict growth of the more fragile shigellae such as Sonnei and S. dysenteriae 1, but they prevent overgrowth of competitive organisms which are commonly found in foods.

The less effective media for isolating Shigellae are Brilliant Green and Bismilth Sulphite agars. MacConkey, Tergitol 7, and

Heart Infusion agars have low selectivity, and Shigellae may be overgrown by competing microflora, but these media will allow the more fragile Shigellae to grow. Xylose Lysine Desoxycholate (XLD) Agar has intermediate selectivity W.I. Tayler, 1965. Incomporative studies, this medium has shown superiority over other media for isolating Shigellae, but S. dysenteriae 1 grows poorly on it. This medium contains Xylose, as a differentiating agent. Since most Shigellae do not ferment Xylose they appear as alkaline (red) colonies on the XLD plate. Shigella strains which ferment xylose rapidly can be missed on this medium, however, therefore a second plating medium containing lactose (which is not fermented by freshly isolated Shigellae) should be used Shigella species undergo colonial variation on usual culture media. Therefore a Shigella strain may occasionally change colonial morphology, and thus it may not be detected.

Method—Plate Count Method

When a sample is frozen, thaw a portion of it for analysis.

Take 25g of sample, cut products such as meat or vegetables into small pieces with sterilized scissors and transfer to a stomachir Add 225ml of Gram-negative Broth and thoroughly mix sample with broth. Incubate at 35—37°c for 18 hours.

Procedure for plating on selective Agar MEDIA

1. Prepare dried plates of three or four selective media, including MacConkey Desoxycholate citrate (DCA) and Xylose lysine, Desoxycholate. XLD agars.

2. Transfer a 5-mm loopful of each enrichment broth culture to the surface of each of three selective agar media and streak in a manner to obtain isolated colonies.

3. Incubate plates inverted at 35-37°c for 24 hrs and then examine colonies.
   (a) Typical Shigellae on XLD Agar appeared or pink colonies usually 1mm in diameter. Colonies with black centres are not Shigella but are likely to be Protous, Salmonella or Arizona.

(b) Typical Shigella colonies on Desoxycholate citrate, MacConkey and SS Agars appear opaque, with ground glass appearance and with even margin.

4. If typical growth has occurred select three suspect colonies from each selective agar medium.

Biochemical screening for Shigella

1. Purify suspect colonies by streaking on separate plates of MacConkey Agar to obtain isolated colonies.
2. Transfer cells from a separated colony to a tube of Tripal Sugar Iron Agar (TSI) by streaking the slant and stabbing the butt.
3. Incubate the cultures overnight at 35-37°C.
4. Discard cultures that do not give reactions typical of Shigella in TSI Agar. Typical reactions are a red slant (alkanline reaction) and a yellow butt (acid, glucose fermentation), with no H_2S (i.e. no blackening of the medium) or gas production. Shigella do not produce gas except for certain biotypes of Shigella flexneri.
5. On suspect Cultures conduct the Indole test (p 58) the Methyl Red Test (p 67) and voges — Proskaner. Positive indole test and Methyl Red tests are confirmed by Agglutination test.

Procedure

1. Dilute and adequately pretest antisera with known test Cultures to ensures reliability of test results with unknown cultures.
2. Using a wax pencil, mark off two sections abouts 1 × 2 cm on the inside of a glass Petridish or on a 5 × 7.5cm glass slide.
3. Remove a portion of the growth from a Nutrient Agar or Triple Sugar Iron Agar Nutrient Agar or Triple Sugar Iron Agar slant and suspend it in 0.5 ml Formalinized Mercuric Iodine saline solution. Place a small amount (1.5mm loopful) of the thick suspension directly on to the dish or slide in the upper part of each marked section.

4. Add a drop of Shigella antiserurn to one section of emulsified culture and mix with a sterile loop or needle.
5. Tilt the mixture in both sections brack and forth for 1 minute and observe against a dark background. A rapid strong agglutination is a positive reaction.
6. Test each culture with antisere groups A, B, C, D and A—D (Escherichia coli Alkalescens-Dispar group). Heat suspension of cultures that appear to be Shigellae, but which agglutinate poorly or not at all, in a water bath at 100°C for 15-30 minutes. After such treatment cool the suspension and test on slide for agglutination.

These Antisere are available from Biologicl production, Izat Nagar Bareilly (UP) India or Swedish Institute for infections diseases control, Solna, Sweden.

Detection of Shigella spp in food with nested Polymerase chain reaction (PCR)

In some advance Microbiological laboratories PCR methods have been developed but a universal media for PCR for all species has not been developed so far.

Presently in Sweden National Food administration Uppsala laboratory a Nested PCR method has developed. The presence of PCR inhibitors in food samples may inhibit amplification and limit the usefulness of PCR techniques. The buoyant density centrifugation method is successful in food as separation of bacteria and food due to their different buoyant densities in Standard Isotonic Percoll medium.

Milk and Milk Products

GENERAL BACTERIOLOGICAL EXAMINATION OF SPECIFIC FOODS

INTRODUCTION: General and specific enumeration give useful indications of the hygiene value of processed food and raw food. It also allow predictions of the potential shelf-life identification of potential health hazards by use of suitable indicators or by direct detection of pathogens.

This chapter deals with various food articles examinations using different media, methods and special techniques. The detail procedures described here may be required for advisory or legislative standardization of the microbial content of food.

The toxins produced by some bacteria and yeast and Mould if not controlled at the initial stage of growth and multiplication effect the health of consumer and some time leads to dangerous diseases like Cancer.

The fresh vegetable and fruits imported or sourced from fields, fertilized with animals manures, washed with high concentration of chlorine have been found ineffective in reducing the total bacterial. Count which include Coliforms, E.Coli and other potential pathogens like Salmonella.

Spices which contain antibacterial and antifungic oils and other antioxidents and natural preservatives may contain surface contaminants and in due course spoil the flavour and colour and becomes unfit for human consumption.

Raw Milk and its products like Cheese should not be consumed as such and sale at outlets be prohibited. Raw milk is often contains pathogens such as E.Coli, Salmonella, Campylobactor, Straphylococcus aureus, streptococcus species, rickettsia such as Coxiella burnetti and some pathogenic viruses. Raw Cow's milk may contain sometime Brucella and Tubercle bacilli when the animals are suffering from Brucellosis and tuberculous mastitis.

In summer season and rainy season many animals suffer from mould attack, so it is advisable to check milk for aflatoxin 'M'

Only pasteurized and heated milk should be used for all purposes.

MILK IN LIQUID FORM

The following statutory tests are described in Indian standards institution (BIS), United Kingdom regulations and ISO specifications:

Untreated milk test: Full Methylene blue test and 1/2 hour Methylene Blue test.

- **Pasteurized milk:** The phosphatase test
- **Ultra Heat Treated (UHT) milk:** The colony count test.
- **Sterilized milk:** The turbidity test. Tests for Tubercle Bacilli and Brucella organisms are also required.

MILK AND MILK PRODUCTS

A general examination should also be done including Plate Count and Coliform Counts and a search for intestinal pathogens. The content of B. cercus and C. Perfringens are important also.

Raw milk drawn aseptically from a healthy udder will contain only a few hundred to a few thousand bacteria per millilitre, mostly of the genus Micrococcus and the udder depththeroid Corynebacterium bovis. In normal non-aseptic milking, many microorganisms are introduced into even high-quality milks from the external environment-from the udder surface, from bovine faeces, soil, bedding, feed and environment of animals. Bacterial counts increase if after milking the milk is kept for some time. If the milk has been chilled soon after milking and retained at refrigeration temperatures, there will be growth and multiplication of psychrotrophic mesophiles such as Psedomonas, and also certain Bacillus species. If a delay occurs before chilling is applied, then the numbers of lactic acid bacteria may increase substantially.

Chilled raw milk sampled at the collection centre should have aerobic mesophilic count (AMC) at 30°C of fewer than 10^4 per ml. and at pasterution plant or creamery should have an AMC at 30°C of fewer than 10^5 per ml. Microscopic examination for Mastitis should be done. After processing to cheque the quality of finished product and hygienic quality of production and storage, total aerobic mesophillic count (AMC) should be done.

EXAMINATION FOR RAW MILK

Methods for detecting Mastitis:

1. Microscopic examination of Breed's smear stained with New man's stain to detect somatic cells and/or bacteria.
   An alternative method to microscopic method:
   (a) an electronic particle counter and (b) ethidium bromide stain followed by automatic cell counting by fluorescence microscopy, both are ISO methods.
2. Mastitis (inflammation of the udder) Causing pathogens like streptococcus which are common, a medium such as Edward's aesculin.
3. Whitside Test:

For rapid diagnosing of mastitis an indirect method is to assess number of cells present in the milk. When these are present in large number the viscosity of the milk increases due to release of nucleic acid by the cells when treated with alkali.

Method

Mix one part of N NaOH with five parts of milk or udder secretion on a glass plate and stir with a glass rod for 5 second.

Note the extent of the viscosity developed if any, An increase in Viscosity occurs with increase in the number of cells. Absence of increased Viscosity is recorded as negative and indicates normal milk.

CULTURAL EXAMINATIONS

The medium used for plate culturing is aesculin crystal violet of concentration 1 in 500000 which permit the growth of streptococci and inhabit any staphylococci. In this medium black colonies of aesculin fermenting organisms (e.g. *Strep - agalactia* and *Strep. dysgalactiae*) produced colourless colonies.

An alternative selective medium for the pyogenic streptococci in general is Petts Colistrin Oxolinic acid blood agar (COBA), this inhabits Gram-negative bacteria and staphylococci.

METHOD OF ISOLATION

With the help of a loop smear 0.01 ml. of milk on the surface of prepared and well dried blood agar plate. Incubate at 37c for 48h examining after 24h and 48h. Count the number of colonies. Similarly examine with other parts of Milk and compare. The presence of any haemolytic Colonies should also be recorded.

The predominant isolated colonies signifies the presence of causative organisms. Record the size and shape of representative colonies, hemolysis if any, pigmentation, Gram reaction and morphology. These observations are sufficient to indicate the probable genus of the organisms, but further confirmatory tests can be considered if required. Carry out slide Coagulation test as described (see page 85) On suspected staphylococci, a positive reaction indicates the presence of Staphylococcus aureus to physiological tests, the precipitin test (see page 87) my be used to determine the lancefields group of the isolate.

When mastitis-causing pathogens are detected and clinical or sub-clinical mastitis is diagnosed, the pathogens can be examined further to determine their sensitivity to antibiotics.

SENSITIVITY TESTS

To know which antibiotic is effective against the organism for treatment of mastitis, a sensitivity test is carried out on a pure culture of the isolate from the mastitis milk.

PROCEDURE: Subculture the isolate into a suitable broth medium (nutrient broth or yeast glucose lemcobroth) and incubate at 37°C for 24hr. Pour 0.1ml of culture on the surface of a previously poured well dried agar plate (nutrient agar for staphylococci and yeast glucose lemoagar for streptococci). Spread the culture over the plate and allow to dry the agar. Place the antibiotic discs on the surface of the agar using a flamed forceps.

Incubate at 37°C for 24hr and record the presence of zones of inhibition around the discs. Inhibition indicates sensitivity of a particular antibiotic which is therefore potentially valuable for mastitis treatment. A large zone around one antibiotic disc does not necessarily mean that it is more effective than another antibiotic which produces a small zone, as many factors are involved in determining the zone size. A simple method of subsequently determining the minimum inhibitory concentration for an antibiotic showing a zone of inhibition around it is the E.Test, which consists of an antibiotic gradient strip that can be placed on a seeded plate.

AEROBIC MESOPHILIC COUNTS

For Plate count the media used are yeast extract milk agar and plate count agar. Incubation period of mesophilic counts is 3days at 30°C. Psychrotrophic counts and thermophilic counts can be determined by incubating at 4.5 and 55°C for 14 and 2days respectively. When low numbers of thermophilics are expected they may be counted by multiple tube technique by using tryptone glucose yeast extract broth, with microscopic examination after incubation. In examining milk from refrigerated bulk tanks, and overnight collection, the psychotropic count becomes of great importance.

Presumptive Coliform and Esch. Coli counts.

Coliforms can enter in the milk from uncleaned equipments and unhygienic place of storage. Tests for Coliforms is important as it gave imperfect and irregular indices of contamination and is statutory requirement Methods to follow as stated at page no 129 to 134.

SANITARY QUALITY OF MILK

Methylene blue or resazurin tests are generally done to assess the quality of fresh non chilled milk. The best method to assess the quality of refrigerated milk stored in tankers are colony counts at 5 and 30°C and theromoduric counts. Aseptically drawn non chilled fresh milk contains 500-1000 bacteria per millilitre. On the basis of counts the milk is further subjected to various treatments for preservation and transportation.

DYE REDUCTION TESTS.

A group of tests which are carried out depends on the response of a number of Redox dyes to the presence of metabolically active micro-organisms. The redox dyes are able to take up electrons from an active biological system and this results in a change of colour. Usually the oxidized form is coloured and the reduced form colourless but the triphenyltetrazolium salts (TTS) are an important exception. TTS and their derivatives are initially colourless and become intensely coloured and usually insoluble after reduction to formazones. This reduction process is irreversible, and the formozane form of the dye is therefore not reoxidized by molecular oxygen as in the case with methylene blue.

The most widely used redox dyes are Methylene blue, Resazurin and Triphenyltertozolium chloride

METHYLENE BLUE TEST

REAGENT: Prepare solution by dissolving one standard tablet (S.D.fine) in 200 Cold sterile, glass-distilled water in a sterile flask ad make up to 1000ml volume, keep the solution in a refrigerator. Transfer aseptically for daily use in a sterile container.

1. Take 10ml marked ground glass join test tube sterilized and transfer thoroughly mixed 10ml milk, Prepare similarly a set of tubes.
2. Add 1 ml of methylene blue solution till last drops in each tube.
3. Close the test-tubes with sterilized glass joint and invert the tubes slowly twice to mix the contents.
4. Within few minutes, place the tubes in a Water-bath at 37.5 ± 0.5°C and note the time. The level of the water in the bath

should be above that of the milk in the tubes and the bath should be fixed with a lid to exclude light.

5. Set up a control tube of similar colour and capacity. Pour 10ml of milk of same fat content and colour and one ml of tap water. Close with sterilized glass stopper and keep for 3 minutes in boiling water, then cool and place in water bath. The control tube will help to determine when decolorization is complete.
6. Examine the tubes after half an hour. The milk is regarded as decolorized when the whole column of milk is completely decolorized or decolorized to within 5mm of the surface. A trace of colour at the bottom of the tube may be ignored provided it does not extend upwards for more than 5mm.
7. In case the colour of the tube have not changed within half an hour period, the tubes should be examined at half hourly intervals for the duration of the test. Tubes that have decolorized should be removed from the water bath, tubes in which decolorization has begun should remain in the water-bath without inversion until decolorization is complete. All other tubes in the water bath should be inverted once to redistribute surface cream within the milk and replaced. Excessive inversion should be avoided because it results in reoxidation of the methylene blue and consequently wrong results expected.

TEST INFLUENCE: Samples are regarded as satisfactory provided the methylene blue is not decolorized after 30 min. at 37-38°C.

RESAZAURIN TEST: The test is generally done when raw milk contains large numbers of Lucocytes and also in Mastitis milk. The reduction of resazaurin dye tubes place at two stages, first irreversibly into resorufin through shades of blue and mauve to pinky and then reversibly from resorufin into the colourless dihydroresorufin. The colours produced can be matched in a Lovibond comparator and given a number ranging from 0 (colourless) through 1 (pink) and 6 (Blue)

1. Resazurin solution: Dissolve one standard tablet (SD fine chem.) weighing 1g in 50ml of cold sterile glass distilled water and make up the volume 200ml to give a .005% solution. The solution is kept in cool dark place, preferably in a refrigerator and discarded when 8 hr old.

2. Transfer 10ml of thoroughly mixed milk in a ground glass sterile test tube having glass stopper with 10ml marking.
3. Add 1 ml of resazurin solution till last drops.
4. Stoppered the tube and mixed the dye and milk by inverting twice.
5. Transfer the tub to a water bath at 37.5 ± -0.5°C and note the time. The water bath should contain a lid to exclude light. Similarly prepare other set of tubes from different samples. Prepare a test tube containing 10ml milk of the same sample, but without the dye (blank). Keep the tubes for 10 minutes. Tests can be carried out for 1hr and 3hr. The 10 minutes Resazurin test is useful and rapid, a screening test used at the milk platform. The 1hr test and 3 hr tests provide more accurate information about the milk quality but after a fairy long time. They are usually carried out in the laboratory.

Place the sample test tube in a lovibond comparator with Resazurin disk (No. 4/9) and compare it colour *metrically* with a test tube containing 10ml milk of the same sample, but without the dye (Blank)

Resazur Indisc. No.	*Colour*	*Grade of Milk*	*Action*
6	Blue	Excellent	Acceptable
5	Light Blue	V. Good	Acceptable
4	Purple	Good	Acceptable
3	Purple Pink	Fair	Separate
2	Light Pink	Poor	Reject
1	Pink	Bad	Reject
0	White	Very Bad	Reject

THERMIZED MILK

Thermization is a process in which raw milk is heated at about 60-65°C for around 15S to reduce the numbers of heat sensitive psychrotrops such as Pseudomonas which could otherwise grow during refrigration of raw milk, producing extracelluar enzymes (e.g. proteases and lipases)that may persist after subsequent heat treatment to give rise to spoilage of dairy products such as UHT treated milk.

PASTEURIZED MILK

Milk is heated to at least 63°c and held at such temperature for at least 30 minutes, or heating it to at least 71.5°c and holding at such temperature contineously for least 15 seconds or an approved temperature-time combination that will serve to give a negative phosphatase test. The Milk shall be cooled immediately to temperature of 10°C or less. Pasteurization in these conditions eliminate Coxiella burnetti, a rickettsia responsible for fever which can be transmitted in milk. Recent disease outbreak attributed to milk and other dairy products have implicated Listeria monocytogens as the causative agent.

Pasteurized milk is tested for Aerobic mesophilic counts at 30-32°c and coliform or total Enterobacteriaceae counts and alkaline phosphatase (APT) test.

Microbial quality of pasterurized is now E.U. based regulations which require pasteruzied milk to contain

Pathogens	absent in 2.5 ml (n = 5, c-0)
Coliforms	Per ml m = 0, M-5, n = 5, c = 1
Aerobic mesophilic	counts (21° c) m = 5 × 10^4,
	M = 5 × 10^5, n = 5, c = 1

Milk after storage at 5°c its counts at 21°c should be less than 10^5 cfu per ml.

Acceptance and rejection of samples Microbiologically is covered under three class—Attributes Planes. A three class plan is defined by four numbers

n= the number of samples to be taken from a lot.

M= A count which if exceeded by any of the test samples would lead to rejection of the lot.

m= A count which separates good quality from marginally acceptable quality and which most test samples should not exceed.

c= the maximum number of test samples which may fall into the marginally acceptable category before the lot is rejected.

UHT (ULTRA HEAT TREATED MILK)

In this process Milk is heated to 135°c for one second. The UK standard for UHT milk and cream is that after incubation in a closed container at 30°c for 15 days and at 55°c for 7 days, the plate count of 30°c should be 100 per ml or less.

STERILIZED MILK: Sterilized milk is a product which is heat processed in bottle at temperature above 100°c for sufficient time typically 20-40 min for it to pass the turbidity test. The test is based on the principal that the heat process is sufficient to denature whey proteins. In it, casein is precipitated with ammonium sulphate and filtered off. The filtrate is then heated, if it remains clear the milk is acceptable because the whey protein has already been denatured and removed with the precipitated casein. Turbidity indicates an inadequate heat process which has left some whey protein undenatured.

Gram-ve-psychrotropes will not survive pasteurization, although some psedomonds produce extracellular lipases and proteases which are heat resistant. If enough of these bacteria are present ($> 10^5$ ml^{-1}) sufficient enzyme can be produced pre-pasteurization to cause rancidity and casein degradation in processed milk.

MILK POWDER

Generally there are four varieties of Milk Powder prepared from fresh buffalo and Cow milk, differentiated on the basis of Fat Contents and sucrose.

(1) Whole Milk Powder (2) Skimmed Milk Powder (3) Partly Skimmed Milk Powder (4) Sweetened Milk Powder

Two processes are used for drying the milk and making powder and are called Roller dried and spray dried. The micro flora of milk powder depends upon the heat treatment given the milk before drying process and the method of drying the milk. A severe exposure to heat occurs in case of Roller dried process, as a result micro flora is restricted. Spray dried process require less heat, the contaminating bacteria and sensitive pathogen may enter during process of milk powder. There is also chance of contamination from plant as well as from milk prior to drying. It is always better that milk should be pasteurized first and care should be taken that there is no possibility of recontamination between pasteurize and drier, the level of Coagulase-positive. Staphylococcus aureus should be checked to prevent production of exterotoxin which creates health hazards. During packing of milk powder there is a chance of yeasts and moulds atmosphere contamination. During storage of dried milk, any surviving or contaminating bacteria will slowly die, but lower the moisture content the better they will be able to survive. The reconstituted milk should not be kept longer than fresh milk.

EXAMINATION OF MILK POWDER:

The sample is drawn with a dry sterile spoon from the top layer, from middle and lower layer of the Container, mix it up and keep in a sterile container. It is more accurate to draw samples from upper layer and lower layer from the Container and analyze separately.

Weigh out aseptically 10g of composite sample of Milk Powder into a 250ml of sterile Conical flask and add sterile distilled water at 50°c with gentle shaking, make up to the 100ml mark. Shake the flask till whole the powder is dissolved and place the reconstituted milk on water bath at 50°c for 15 min. This is 10^{-1} dilution. Prepare further dilutions in 9 ml amounts of 0.1% peptone water as required or in Ringer's solution. Some standard methods specify the use of 0.1% peptone or even 0.1% peptone with 0.85% sodium chloride for reconstitution of initial dilution. It is observed that use of sterile distilled water gives better results in initial reconstitution due to the fact that some microorganisms metabolically damaged by exposure to hyper tonic solutions in the 10^{-1} solution.

MICROSCOPIC EXAMINATION: Take one sterile (flame heated) glass slide (2 × 2.5 cm) and spread 0.05ml of milk from lst/2nd dilution. Stained with Newman's stain and examined microscopically. The average number of bacteria or cell per field is determined and calculate the total number of bacteria or cells per gms of the sample. The count gives an indication of the extent to which microorganisms may have proliferated in the milk before drying.

CULTUREAL EXAMINATIONS

Aerobic mesophilic counts: Transfer 1 ml of each dilution in sterlized plates for the following tests:

1. Aerobic mesophilic cunts. Use plate count agar or yeast extract milk agar, incubate at 30, 37 and 55°c for 5, 3 and 3 days respectively.
2. Coliform or 'total Enterobacteriaceae count. Use MPN method, for the 'Total Enterobacteriaceae' Count. The medium used is lauryl sulphate trptose broth containing glucose and incubating at 30°c. For Coliform Count use lauryl sulphate tryptose broth containing Lactose in MPN method incubating at 30°c.
3. Enterococci: Use plate counts method and the media used in either Packer's crystal violet azide blood agar or maltose azide agar. Detail given under the heading Enterococcus page 138.
4. Yeast and Moulds: See Yeasts and Moulds section.
5. Staphylococcus aureus: See Plate count method use Baird (page 166) Parker's medium, incubate at 37°c for 24h when microscopic count found very high (e.g. more than 5×10^5 cocci) it is desirable to test for Enterotoxin by ELISA. See page 170.
6. Bacillus spores and B.Cereus: See under heading Bacillus as indicator organisms.
7. Clostridium Spores use plate count agar or blood agar (the latter as surface counts) incubate at 30, 37, 55°c anaerobically for 3-5days Clostridium perfringen can be enumerated by a multiple tube method as described under its heading. Page 159
8. Salmonella. Method described under detection of Salmonella. Use reconstituted milk(10^{-1} dilution incubated at 37°c for 24h to 48h). See page 148.

Recommended specifications

Microscopic Count (Spray dried milk Powder):	Less than 10^6 per gram
General viable Count	Less than 10^4 per gram
Counts of Coli-aerogenes bacteria and Yeast's & Moulds	Less than 10 per gram
Staphylococcus aureus	Less than 10 per gram
Salmonella	Absent from 100gm of sample

For Roller dried milk Powder	
General viable count	Less than 10^3 per gram
Counts of Coli-aerogenes bacteria and Yeast & Moulds	Less than 10 per gram

The UK standards for dried Milk and dried milk based products:-

Salmonella	
Dried milk	absent in 25g, n=10, c=0
Dried milk products	absents in 25g, n=5, c=0
Listeria monocytogenss	absent in 1g
Staphylococcus aureus (per gm)	m=10, M=100, n=5, c=2
Coliforms at 30deg Celsius (per gm)	m=0, M=10, n=5, c=2

Evaporated Milk unsweetened Canned:-

There are chances of contamination of milk during canning either due to seams leaking or cans not properly sterilized prior to packing.

EXAMINATION: Incubation Test: Incubate representative Cans at 55°c for 7 days, 37°c for 14days and 25-27°c for 1 month. Examine the cans after the stipulated incubation period. On incubation cans may become swollen if gas producing micro-organisms are present in the milk. Some times on incubation cans become flipper if not properly exhausted at the time of seaming. Hard blown cans clearly indicate that milk is not microbiologically sound.

CUT OUT EXAMINATION: Incubated cans and un-incubated cans should be examined separately Cleans the cans with towel, remove the labels and wash the cans with warm water. Wipe the can dry with tissue paper, clean the can with cotton swab moistened with ethonal and flame the surface for few seconds. Make the opening with sterile can opener and take sample with sterile 1ml pipette and transfer into crossley's milk peptone medium, 1ml to 10ml molten yeast glucose lemco agar for a shake tube, 1ml into a Petridish and pour with yeast milk agar or plate count agar. Incubate at temperature as stated in incubation test.

For Microscopic examination, prepare a smear of the milk on a slide and stain by Gram's method, defatting first with xylene if necessary, No organism found in 1ml may be assumed that milk has been sterilized.

Condensed milk Sweetened Canned: It contains about 40 to 50% of sucrose, low available water (aw) which restrict the multiplication of any microorganisms present in the product. Cans are not sterilized due to presence of high content of sugar and caramalized on sterilization. Yeasts and moulds may grow in this condition. These generally grow when cans are under filled with large head space, providing a source of oxygen for mould growing on the surface of the product. Some time 'buttons' formation of compacted mycelium is seen on the surface of product.

CUTOUT EXAMINATION

1. Warm the Can in Water bath at 45°c for 10-15 min to reduce viscosity. Prepare the Cans for cut out examination as stated in Evaporated Milk.
2. Cultural Examination: Weigh out a sterilized flask containing 90 ml of diluent. Keep the flask and Can near the flame and transfer 10 gm of milk with sterile spoon or with sterile 10ml pipette with large orifice Prepare further dilutions up to 10^{-4} as described in the method of preparing dilutions.

Transfer 1ml of each dilution to sterile plates and use suitable media as follows:

(i) Aerobic mesophilic Count: Use yeast extract milk agar or plate count agar and incubate at 30°c for 3 days.

(ii) Lipolytic mesophilic count: Use yeast extract milk agar or plate count agar and incubate at 30°c for 3 days (see Hydrolysis of tributyrin)

(iii) Coliform Counts: See under Coliform organisms and E.Coli. Use violet red bile glucose agar or violet red bile lactose agar respectively and incubate at 30°c for 24hr. To detect few numbes Use a multiple tube counting method. Presumptive positive results from multiple tube MPN counts transfer a loopful from a positive tubes into tubes of similar medium or by streaking on MacConkey's agar.

(iv) Yeasts and moulds count: see under chapter Yeasts and Moulds.

(v) Report the results per gram of condensed milk.

SPECIFICATION LIMITS

General Viable counts	Less than 100 per gram
Lipolytic Counts	Less than 10 per gram
Yeasts and Moulds Counts	Less than 1 per gram
Coliform Counts	Less than 1 per gram

CREAM: Cream means the product of Cow or buffalo milk or a combination thereof. Three catagories of cream on the basis of Fat Contents are made e.g. 25, 40, 60 % by Wt.

In international market cream is manufactured from fresh pasteurized milk and cooled to below 5 degrees C. The temperature of cream must at no time exceed 7-5 degrees C when stored for day or so,

The methods of cream examination are similar to that of milk.

METHOD: Sample is taken from Can/bottle/carton with a sterile spoon and collect in a sterile Jar. The sample should be examined within 2 hrs. of production or otherwise keep in the refreigerator until testing begins.

Cultural examinations: Prepare decimal dilutions in sterile distilled water up to 10^{-4}. Transfer 1 ml of decimal solution from each dilution to sterilized plates for the following cultural examinations:

(i) Yeast extract milk agar or plate count agar for aerobic mesophilic counts incurbated at 30°c for 3 days.

(ii) Use VRBA with glucose and VRBA with Lactose for total Enterobacteriaceae and Coliform counts respectively. Incubate at 30°c.

(iii) Use Baird- Parker's medium for Staph.aureus(surface counts) at 37°c. (iv) Tributyrin agar incubated at 4-5°c and 30°c for 14days and 3days respectively for lipolytic psychrotrophs and for a general lipolytic count.

(v) Examination for presence of Salmonella(see under Salmonella)

(vi) Examination for the presence of Listeria monocytogenes (see under detection monocytogenes)

BACTERIOLOGICAL DATA

	Specification Limit
Total Count/gm	10,000 maximum
Coliforms/gm	-ve
Salmonella/100gm	-ve
Enterobacteria/gm	<10
Yeasts & Moulds/gm	<10
Listeria monocytogenes	-ve

STARTER CULTURES

Starter cultures are those microorganisms that are used in the production of cultured dairy products such as Yogurt and Cheese. The natural microflora of the milk is either inefficient, uncountrollable and unpredictable, or is destroyed altogether by the heat treatment given to the milk. A starter culture can provide particular characteristics in a more controlled and predictable fermentation. The primary function of lactic starter is the production of lactic acid from lactose and other functions of starter cultures may include the following.

Flavour, aroma, and alcohol production, proteolytic and Lipolytic activities and inhibition of undesirable organisms. There are two groups of lactic starter cultures. Simple or defined. Single strain, or more than one in which the number is known mixed or compounded, more than one strain each providing its own specific characteristics. Starter cultures may be categorized as mesophilic or thermophilic.

Lactococcus lactis subsp.cremoris
L.delbrueckii subsp.lactis.
L.lactis subsp.lactis biovar diacelactis
Leuconostic mesenteroides subsp.cremoris.

Thermophilic Starters are Streptococcus salivarius Subsp.thermophilus (S.thermophilus)

Lactobacillus dellerueckii subsp. Bulgaricus.
L. delbrueckii subsp.lactis.
L. Casei
L. helveticus
L. plantarum

Mixtures of mesophilic and thermophilic microorganisms can also be used as in the production of some cheeses.

BACTERIOPHAGE

Bacteriophages are Viruses that require bacteria host cells for growth and reproduction. Initially, the bacteriophage attaches itself to the bacteria cell wall and injects nuclear substance into the cell. Inside the cell, the nuclear substance produces shells or phage coats for a new bacteriophage which are quickly filled with nucleic acid. The bacterial cell ruptures and dies as the new bacteriophage are released.

Bacteriophages are of most concern in cheese making. They attack and destroy most of lactic acid bacteria which prevents normal ripening known as slow or dead vat.

Starter culture preparation: In these days commercial manufacturers provide starter cultures in lyophilized (freeze-dried) or liquid nitrogen-frozen in ring-pull cans. These can be used for preparation of bulk starter

Microscopic Examination of starter culture:

Gram's staining method is used. as described in staining method.

Defatting with xylene is not required for the examination of skim milk cultures. See any difference in relative numbers of pairs of Cocci and short or long chains of Cocci. In starter cultures long chains of Cocci are indicative of Lactoc.Cremoris.

ACTIVITY TEST : Take 250mls capacity sterile flask, add 100ml of sterile milk and 1ml of starter culture, incubate at 30°c in a water bath for 6 hr. After incubation determine the acidity that has developed. Take 10ml of the cultures, add 1ml of 0.5% phenolphthalein and titrate with N/9 sodium hydroxide until a faint color appears. Divide the number of milliliters of N/9 NaoH used by 10ml of culture. The net result is the titratable acidity of the culture.

Activity test for detection of phage in cheeseway

Take two 250 ml flasks containing 100ml of sterile separated milk and inoculate with 1ml starter culture. To one of the flasks add 1ml of 0.1% of whey, to the other add 1.0ml of diluent to act as a control.

Incubate in a water bath at 30deg Celsius for 6hr. Calculate the titrable acidity of both the cultures as started in activity test. If the activity of the culture with added whey is more than 10% less that of the control, it may be considered that the whey contains phage specifically affecting that particular starter culture.

BIOCHEMICAL TESTS FOR STARTER

Catalase test : To 5 ml of starter culture add 1 ml of 10 vol. hydrogen peroxide in a test tube. If effervescence take place it indicates that the starter is contaminated.

Voges Proskauer test: See under section involving carbohydrates reactions - page—67.

Take 2ml of starter culture in a test tube, add V-P reagents and shake the tube well. A pink coloration developing within 30min constitutes a positive reaction indicates that the starter is contaminated.

Cultural Examination

Plate count method for viable starter bacteria prepare decimal solutions of the starter culture in the diluent up to 10^{-10} or as per Microscopic count indication. Take 1ml of the each of the last three dilutions and transfer the plates and pour media as stated below:-

1. Aerobic mesophilic count:-use yeast lemo agar and incubate at 30deg Celsius for 3days.
2. For citrate-fermenting organisms Leuconostoc, Lacto, Lactis subsp etc. use the tomato juice lactate agar and incubate at 21deg Celsius for 4 days.
3. For differentiations of Lactoc, lactis subsp, lactis and Lactoc, lactis subsp, cremoris, use arginine tetrazolium agar. Incubate at 30deg Celsius for 24hr-48hr. Colonies of Lactoc, lactis subsp, lactic are red (arginine), where as colonies of Lacto, lactis subsp, cremoris are white (arginine). The percentage viability can be determined by comparing the colony count with the Breed's semar count (in which a clump or chain is taken to be the equivalent of one colony forming unit).

Method of checking impurities in culture

To check the purity of culture, the presence of contaminating organisms is done by inoculating suitable selective media with 0.1ml of starter culture. For quantitative estimations prepare decimal solutions of the culture and plate 1ml of each dilution on the following media.

1. For yeasts and moulds detection/count use media as described under the heading "yeasts and moulds" page 89.
2. For total Enterobacteriaceae or coliform detection/count use VRBGA or VRBLA respectively or use MPN count method.
3. To detect the presence of non lactic acid bacteria use nutrient agar plates and incubate aerobically and anaerobically at 30deg Celsius for 3 days.
4. To detect citrate fermenting organisms in cultures of Lactoc. Lactis subsp. Lactis and/or lactoc, Lactis subsp. cremoris use semisolid citrate milk agar production of gas will show the fermentation of citrate. The citrated milk is prepared by adding 0.5ml of 10% sodium citrate solution to 10ml of milk. Inoculate citrated milk with 1% (0.1ml) of starter culture and add to 4ml of molten 2% agar at 48deg Celsius. Mix the contents and incubate at 30deg Celsius for 3 days. The presence of organisms will produce carbon dioxide during fermentation.
5. Contamination due to pathogens like staph aureus, Listeria monocytogenes, salmonella etc. may be detected as described under detection of pathogens.

PURIFICATION OF STARTER CULTURES

From colony count plates as prepared in the method stated above, transfer suitable colonies to yeast glucose lemco broth or litmus milk and incubate at 30deg Celsius for subculturing. For the growth of Leucomostoc spp. the broth medium should be used. The culture obtained may be re-examine and restreak.

Preservation of pure culture can be obtained by inoculation into yeast glucose chalk litmus milk and incubating at 30deg Celsius for 24hr or until a slight acidity develops. Robertson's cooked meat medium may be used. Store the culture at room temperature and subculture every 3-6 months.

DETECTION OF RAW MILK IN PASTEURIZED MILK BY FLUOROPHOS

Pasteurized milk may contains raw milk which can pass mycobacterium Para tuberculosis (the organism responsible for TB) in it. The Alkaline phosphatase test (ALP) is done to establish that milk has been correctly pasteurized.

The test involves measurement of the amount of ALP remaining "undenatured" after pasteurization. Reaction with a substrate is then used to facilitate ALP measurement. The chemistry is common to all the ALP tests Residual ALP breaks the bond between a phosphate group and the remainder of the molecule in a phosphate monoesters have been investigated, all liberating different compounds for subsequent measured.

Two methods have been standardized internationally to form statutory approved tests for pasteurization. One, used in the United states (scharer test) is based on the liberation and measurement of phenol and a second, used in the united kingdom (Aschaffenberg & Mullen) is based on the liberation and measurement of p-nitro phenol.

The lowest level of sensitivity was, and still is, 10Lg of p-nitro phenol. Since raw cow's milk & Buffalo milk contains about 10,000 (g p- nitro phenol unit of ALP).

The best colorimetric methods used for over 70 years are able to measure about 0.1 percent raw milk. The Fluorophos which is based on the same chemistry and involves liberation of chemicals measured by fluorescence rather than color, it is more sensitive, more rapid and more reproducible than older test methods. The Flurophos is internationally approved (ISO, IDF, AOAC), is capable of measuring down to 0.003% raw milk. Correctly pasteurized milk in practice yields values of 20-50mu/L.

Fluorophos instrument is manufactured by "advance Instruments inc. USA and available in India and other countries.

DETECTION OF ANTIBIOTICS IN MILK

Antibiotics are used for the prevention and the treatment of cows & buffalos having bacterial infections of the udder (mastitis).

Residue of antibiotics will occur in the milk from the treated animals. The use of antibiotics should be strictly observed to present contaminated milk reaching the consumer. Milk is examined for antibiotics residues because they pose a hazard to consumers, possibly causing allergic reactions and the development of bacterial resistance. It has also an inhibitory effect on starter cultures of bacteria used to make fermented milk products.

Recently two rapid methods have been developed for detection of antibiotics residue in milk.

1. Eclipse test:-commercial microtiter plates of Eclipse are inoculated with 60ml milk and with known concentration of antibiotics in 60ml milk. Incubate for 3 hr at 65 degree Celsius. Measure the sensitivity by Eclipse Instrument. Antibiotics up to 2 ppb residue in milk can be measured.
2. Biosensor test:-A prototype of sensor system for the detection of antibiotics comprises a laser at 635nm wavelength, a moving mirror attached to a flexible mechanical structure, which changes the angle of incidence of the laser beam without changing the incidence position on the waveguide. Fluorescents light is detected with a PIN silicon photodiode and a lock - in amplifier and a signal is recorded as a function of mirror position. Software controls the instrument and display the result.

FERMENTED MILKS

Milk ferments when sufficient numbers of lactic acid bacteria are present. Fermentation occurs naturally by the development of bacteria already present in the milk and as a results milk becomes sour due to Lactobacillus acidophilus and lactic streptococci high growth. Now a days fermented milks are produced by introducing Starter cultures. Different countries make different fermented milk products such as curd in India, yammer in Denmark, Kefir in Eastern Europe and yogurt in Bulgaria and now popular in most part of the world. The consistency, flavour and aroma of these products very from one region to another, and due to different types starter cultures e.g. curd is produced using Lactobacillus acidophilus, yogurt by using Streptococcus Salivarius subsp thermopiles (ST) and Lactobacillus del brueckli subsp. Bulgaricus (LB).

The presence of high number of contaminating yeasts in these products may cause a problem of off flavor during storage.

Examination of culture organisms and contamination:

Microscopic examination: Take a clean slide, put a drop of water and prepare a smear of fermented milk and dry it. When milk products under tests contains high fat, first defat the smear by flooding with xylene, draining and drying before staining. Follow Gram's method of staining.

This gives the number of microorganisms present in the fermented product. Contaminants will be detected only when they are present in considerable number. For detecting low numbers cultural methods are used.

Detection of micro flora in fermented milk

Generally fermented milks are produced by using mesophilic species of lactic acid bacteria such as Lactococcus lactis. The media used for quantities determination is neutral red chalk lactose agar and for Lactobacillus MRS agar. For differentiation of Lactoc, lactis subsp. cremoris and Lactoc, lactis subsp. lactis can be differentiated using arginine tetrazolium agar.

Lactobacillus spp: Medium Rogues agar is used for isolation of Lactobacilli and for particular species of Lactobacillus temperature of incubation is different. Streptobacteria are lacto-bacilli (e.g. Lb easei) are grown at 30deg Celsius and Lb.acidophilus and Lb. Bulgarians are grown at 37deg Celsius. LS differential medium is used to detect Lb. Bulgarians in Yogurt, and to differentiate this organism from Streptococcus thermophilus observe the isolated colonies under microscope. Subculture from Gram-positive rods into litmus milk and incubate at 30deg Celsius and 37deg Celsius as stated above. Re-examine the culture when growth is obtained.

LACTOCOCCUS AND STREPTOCOCCUS

Prepare decimal solutions of the substrate and inoculate 1ml into neutral red chalk lactose agar as described in plate count method. Incubate at 30deg Celsius to isolate lactococci from sour milk and from cultured buttermilk, and at 37-43deg Celsius to isolate strep thermophillus from Yogurt. L-S. differential media can also be used to isolate Strep thermophillus.

For sour raw milk the differential media with addition of thallium acetate in the final concentration of 1:2000 may be made selective for Gram-positive bacteria and also helps to suppress the coliforms bacteria.

On neutral red chalk lactose agar, acid producing colonies of lactococci and Streptococci are of deep red colour, small in size and surrounded by clear zone.

Select isolated presumptive lactococci and streptococci colonies for microscopic examination and also confirm the presence of Gram- positive cocci by Gram's method. Inoculate the litmus milk with selected colonies and incubate the media at 30deg Celsius or 37deg Celsius until acid is produced, then re-examine. Obtain a pure culture by restreaking on a solid medium and confirm as stated above.

YEASTS

Detection of yeasts in the fermented products is important as high count may spoil these products. Use the selective medium for growth of yeasts, the media are Davis yeast salt agar. Potato dextrose agar and Dichloran rose Bengal chloramphenicol agar. For isolation, picked off the colonies from these media plates and transfer to malt extract agar slopes. For confirmation, inoculate growth into Lactose, glucose and galactose broths and look for fermentation.

BUTTER

Home made butter is made from Buffalo milk and cow milk. Milk is usually heated to 95-100deg Celsius and cooled to room temperature, (20-25°c). A pinch of curd is added and kept for overnight for fermentation. The milk changes to curd. The curd is diluted, churned to give Butter and Butter milk and to give butter and buttermilk (Lassi or chaas). Commercially the butter is made from cream. Fresh milk is pasteurized, cream is separated by a cream separator (30 to 40 percent extraction), aged overnight at 5 to 10°c and then ripened or fermented for 15 to 16hrs, after adding starter culture. The culture most frequently used is

Streptococcus lactic, perhaps also spp cremoris. 1 to 2% starter culture is used to convert the lactose in milk to lactic acid. The ripened cream is next rotated in a churner which separates globules of fat into large, pea sized clusters of butter and separate out from watery portion. The water is drained off. Butter may be either salted or unsalted. Salt is added between 0.5 to 2.5% and moisture in finished product is 16% as per PFA and other countries leg islation. Presence of salt restricts the growth of many organisms ICMSF(1980b) has described two outbreaks of staphylococcal food poisoning found to be caused by whipped butter. Storage of butter at low temperature retards microbial growth. Spoilage of butter is therefore most likely to arise from activity of microorganisms capable of growing at low temperature, especially those capable of lipolysis, proteolgsis, off flavour or those causing discoloration.

The quality of butter from hygienic point of view can be assessed by aerobic mesophilic count at 4-5° and 30°c, yeast and mould count, lipolytic count, halotolerant and Coliform count. There is rare chance of Staph.aureus and Staphylococcal enterotoxin.

PREPARATION OF SAMPLE

Procedure to draw sample has been described in ISO 707:1997. In practice sample from butter paper packed packets is drawn from upper layer of 5mm from different packets. The sample is taken with a sterile knife and collected in a sterile sample jar aseptically. Similarly a sample should drawn from the deep layers and collected as describe above. Samples from bulk containers of different sizes are also drawn from upper layer and from middle layers separately. A sample of at least 50g should be taken. It has been observed that there are chance of the upper surface of the butter may contain yeast and mould growth due to contamination from butter paper used. Examination of both samples should be done separately to assess the quality of butter.

METHOD FOR CULTURAL EXAMINATIONS

The composite sample drawn in a Jar be immersed in Water-bath at 45°c. Mix the melted sample by shaking several times in water bath. Canned Butter be melted similarly after clearing and removing wrapper. Pipette out 10ml of melted butter sample and transfer to serile 250ml flask containing 90ml of 0.10% peptone with 0.10%agar to stablize the emulsion. The pipette and diluent should be kept at 45°c. Prepare decimal solutions up to 10^{-4} or as required.

Examine for the following parameters.

(1) Total Plate Count: Use plate count agar and incubate at 30 and 4-5°c fr 3 and 14days respectively.
(2) Counts of Yeast and Moulds as described under Yeasts and Moulds Count page 87. 89
(3) Counts of casein hydrolysis organisms:- Method described under Hydrolysis of casein. Use 30% milk agar, incubate at 30°c for 5 days. Caseolytic colonies are surrounded by clear zones and may be confirmed by flooding the plates with a protein precipitant before counting the colonies.
(4) Lipolatic Count: Generally tributyrin agar is used for lipolatic count, other media's described under heading reactions involving Lipids. Phospholipds and related substances are also used. Since Butter contains more complex fats, tributyrin agar in addition to tributyrin agar other medias are used.
(5) For total Enterobacteriaceae and Coliform bacteria violet red bile glucose agar or VRBLA is used and incubate at 30°c for 24hrs. MPN method may be used.
(6) For Halophilic organisms count use 15% Sodium Chloride to prepare dilutions in place of peptone water and media containing 15% Sodium Chloride.

SPECIFICATION OF BUTTER

(U.K.standard.)

Aerobic mesophilic count	< 10000 per gram
Count Protolytic, lipolytic & psychrotropic counts	< 1000 per gram
Yeasts and Moulds Counts	< 100 per gram
Coliform Counts	< 10 per gram, m=0, M=10, n=5,c=2
Salmonella	Absent in 25g, n=5, c=0
L.monocytogenes	absent in 1g

CHEESE: Cheese is made from Buffallo/Cow Milk raw or pasturized by treating with starter culture of Streptococci or Lactobacilli. The role of starter organisms is to change the lactose present in the milk to lactic acid, these strains of microorganisms also greatly affect the flavour of the final product. Rennet is added after 45 minutes of starter culture. Rennet is used to catalyze the conversion of casein in milk to para-casein by removing a glycopeptide from the soluble casein. Para-casein further clots i.e Coagulates in the presence of Calcium ions to form White, Creamy lumps called the Curd, leaving behind the supernatant called whey. After the Curd is separated from the whey, salt, seasoning, and other curing and flavouring ingredients are added. The Curd is wrapped in cheese cloth and pressed for 12 to 18 hrs to remove the additional whey soaked in the curd. The curd hardens and forms a cheese block in the shape of the press as the whey is squeezed out. Finally, the cheese block is dried for 6 hrs. It is now ready for consumption. Well ripened cheese flavours attract all types of persons. No wedding party in India is without cheese dishes and it looks like a garden without flowers. The problem is the use of Rannet which is obtained from the stomach or abomeson of suckling calves, lambs or goats. Its most important component is proteolytic enzyme renin or chymosin. In India it is called non-veg.cheese. For making vegetarian cheese, vegetables rennet or Microbial rennet is used. Vegetable rennet usually means the enzyme is plant based. The plants like lady's bedstraw, stinging nettle, figleaves, melon,safflower and wild thistle.

Microbial rennet is now produced using the bacterium. E.Coli and Yeasts. This type of rennet can be used to make cheddar or hard cheese.

Microbiological spoilage of cheese depends on the type of cheese, e.g. Soft, fresh, hard and semihard cheese. Spoilage of cheese also depends on water activity aw or and pH maintained at the time of preparation, when these conditions are not properly maintained, most moulds grow on the surface or in crevices and coloured colonies appear on surface. Geotrichum called the dairy mould grows on soft cheese and during ripening some time suppresses moulds as well as surface ripening bacteria. All types of cheese, even hard cheese have been involved in the outbreaks of Salmonella enteritis. In many countries food poisoning cases have been reported due to consumption of soft and semisoft cheese.

The caustive organisms found were staphylococcal, Listeria monocytogenes and Escherichia Coli 0157. In hard cheese Brucella abortus have been reported. Indian Cheese is a soft cheese and gets slimy in short-period even after keeping in a fridge

Examination of Cheese:

Sampling: Sample of cheese should be drawn from different portion of bricks with sterile knief and a composite sample is made in a sterile container. From a cheese tin, sample is drawn from drained out solid mass from different portions. The tin should be kept near flame while drawing sample.

Minimum 200gm of sample should be drawn as per PFA rules. Standard procedures are described in ISO707:1997. *Preparation and dilution of homogenate*: Cut the cheese into small pieces and then mince the sample aseptically with sterile mincer and transfer 10gm cheese into a sterile flask containing 90 mls quarter-strength Ringer's solution Emulsify by shaking with hand or by a stomacher.

Further dilutions can be prepared from above dilutions (10^{-1}) in 0.10% peptone water as described in the method of preparing dilutions.

Microscopic Examinations. Take two clean slides, place a flat cut piece of cheese in between two slides, press the slides, then separate and remove excess cheese with the edge of a slide. Defat cheese with xylene for 1 min and dry in air. Stain by Gram's method & examine the number of fields and record the numbers and kind of microorganisms present.

For quantitative examination use dilutions of the cheese sample and count the organisms by Breed's smear method by using charlett's imporved New man's stain.

Procedure for cultural Examinations

1. Total Enterobacteriaceae, Coliform bacteria and E.Coli and E.Coli for Total Enterdeacteriaceae and coliforms 0157.

Use VRBLA or VRBGA or use MPN method, incubating at 30°c. For E.Coli 0157 see page 137.
For E.Coli see method under E.Coli determination. Page 135.

2. Aerobic mesophilic Count: The media used for enumeration is yeast glucose lemco agar. This media will support the growth of most lactic acid bacteria. This media will suppress Lactobaclli. Incubate at 30°c for 3days.
3. Lactobacilli Count: use pour layer plate method and media modified Rogosa agar. This media suppresses the growth of Lactococci but some time Leuconostocs and pediococci may grow in this media when late cheese ripening is done. Incubate at 30°c for 3 to 5 days.
4. Yeasts and moulds count: See under Yeast and mould chapter,
5. Staph aureus: Baird-Parker's, medium is used and incubate at 37°c for 24 hrs.
6. Clostridium perfringens-see chapter—page 159—161
7. Listeria monocytogenes- see chapter—page 164

The Microbiological specifications: In India PFA has not laid down any specification for cheese. U.K. Standards for Cheeses are:-

1. Hard Cheese produced from raw or thermized milk

Steph-aureus	m-10^3, M=10^4, n=5. c=2
E-Coli	m=10^4. M =10^5, n=5, c=2
L.Monocytogens	Absent in 1g
Salmonella	Absent in 25g, n=5, c=0

2. Hard Cheese produced from pasteurized milk:

L.monocytogens	absent in 1g
Salmonella	absent in 25g, n=5, c=0

3. Soft Cheese produced from raw or thermized milk

Staph-aurus	m=10^3, M=10^4, n=5, c=2
E.Coli	m=10^4, M=10^5, n-5, c=2
L.monocytogens	absent in 25g, n=5, c=0
Salmonella	absent in 25g, n=5, c=2

4. Soft Cheese produced from pasteurized milk.

Staph-aureus	$m=10^3$, $M=10^4$, $n=5$, $c=2$
Coliforms	$m=10^4$, $M=10^3$, $n=5$, $c=2$
L.Monocytogens	absent in 25g, $n=5$, $c=0$
Salmonella	absent in 25, $n=5$, $c=0$

5. Fresh Cheese:

Staph-aureus	$m=10$ $M=10^2$, $n=5$, $c=2$
L.monocytogens	absent in 25g, $n=5$, $c=0$
Salmonella	absent in 25g, $n=5$, $c=0$

DAHI, YOGURT, CHHANNA OR PANEER AND KHOA

Dahi is prepared from boiled milk, cooled to 35-37°c, Starter culture of Streptococci or Lactobacilli is added. In India Dahi is made in homes by seeding boiled milk cooled to 35-37°c with 2.5%(one table spoon to a litre) of good curd from the previous batch, followed by undistributed fermentation and setting in a warm place overnight. Home made Dahi contains mixed organisms and must be consumed in a day or so.

YOGHURT

In Advanced countries, commercial Yogurt has in the last two decades become an immensely popular food. It is a modified form of Dahi. In place of the mixed organisms present in Dahi, a starter culture for yoghurt blend of Streptococcus Salivarius Subsp, thermophilius (ST) and Lactobacillus delbrueckii subsp. Bulgaricus (LB) is added to whole partly-defatted, homogenized milk at 43 to 44°c, stirred gently, and filled in plastic or waxed Cardboard tubes. These are then held for fermentation at 41 to 42°c for 3 hours, followed by cooling for 8 hours at 5 to 7°c. This is a plain yogurt which may be modified in taste and flavour by adding flavours, colour, sugar and fruits to the milk before its incubation.

CHHANNA OR PANEER

It is processed from Cow or Buffalo milk or a combination there of by precipitation with sour milk, lactic acid or citric acid. Fresh milk is heated to a high temperature of 80°c, slow stirred, and repidly coagulated by adding to it a 1 to 2 percent solution of lactic or citric acid or even natural sour whey from a previous operation. The whey is drained out while hot, and the product suspended in muslin cloth bag in a bath or cold water to leach out adhering whey. Thereafter it is run through a kneading machine to give it smoothness. This product is used for making sweetmeats like Rasogolla and Sandesh.

KHOA

Khoa or Mawa is prepared from Buffalo or Cow's milk or cómbination thereof by rapid evaporating the milk to a doughy consistency is an open pan with constant stirring and scrapping with a metal Spatula. The mass is kneaded and fashioned into balls. This is used for making various sweetmeats like Gulab-jamuns, Kalakand etc.

Microbiological examination: All these products contain lactic acid bacteria, with time lapse these increases the acidity of the products and chances of yeasts and moulds to grow are certain.

Microscopic examination: as explained under cheese.

Cultural Examination: Preparation of dilutions: Prepare decimal solutions as explained under Cheese and proceed for determination of Aerobic mesophilic count, Lactobacilli, yeasts and moulds, staphylococcus aureus, clostridium and Listeria monocytogenes, and use the same methods as in Cheese.

Results and recommended standards

It is recommended that these products should not contain aerobic mesophilic count more than 10000 per gram, yeasts and moulds count not more than 100 per gram, Salmonella and L.monocytogenes absent in 25g of sample.

OTHER FERMENTED MILK BEVERAGES

Cultured Buttermilk: This product is made from Skim or whole milk by using culture of S.Lactis, perhaps also spp.cremosis. Milk is heated to 95°c and cooled to 20-25°c before the addition of starter culture Starter is added at 1-2% and the fermentation is allowed to proceed for 16-20 hrs. to an acidity of 0.9% lactic acid. This product is frequently used as an Ingredient in baking industry

ACIDOPHILUS MILK

Acidophilus milk is a traditional milk fermented with Lactobacillus acidophilus(LA) which has been thought to have theraptic benefits in the gastro intestinal tract. Skim or whole milk may be used. The milk is heated to high temperature, e.g. 95°c for I hr. to reduce the microbial load and favour the slow growing LA culture. Milk is inoculated at a level of 2-5% and incubated at 37°c until Coagulated. Some acidophilus milk has an acidity as high as 1% lactic acid, but for therapeutic purposes 0.6-0,7% is more common. Another variation has been the introduction of a sweet acidophilus milk, one in which the LA culture has been added but there has been no incubation. It is thought that the culture will reach the GI tract where its therapeutic effects will be realized, but the milk has no fermented qualities thus delivering the benefits without high acidity and flavours, considered undesirable by some people.

MILTONE.

It is a blend of milk with a slurry of Protein isolated from groundnut oil cakes(ground nut cake made from ground nuts after taking the oil from it) The Oil-cake is dispersed in alkaline Water, filtered to remove insoluble matter. This alkaline water is blended with glucose, vitamins, minerals, flavours and an equal quantity fo whole milk to give a product which resemble milk in looks , taste and nutritive quality. Actually it is a method of toning milk not with milk powder but with vegetable protein. This product is available commercially in flavoured form as a drink in India (Bombay). In mass scale it is used in the supplementary children's feeding programmes which operate in several states in India.

CHAR SATHI

It is a product made by blending skim milk with vegetable fat, milk and vegetable proteins, emulsifiers, stabilizers and Vitamin A, followed by homogenization, It has many qualities demanded in an Indian home, a milky taste, ability to white tea and coffee, and capacity to set to curds, besides being less expensive than milk.

Microbiological Examination: Take a clean slide with a drop of water, smear these products on it, defat it with xylene, drain and dry before staining. Stain the smear by Gram's method. Examine the kinds and relative proportions of microorganisms present in both the products.

Cultural Examinations: Transfer 10ml of each product to 90ml of quarter strength ringer solution to give a homogeneous suspension, the 10^{-1} dilution. Prepare subsequent decimal dilutions up to 10^{-4} or as required.

1. Examine for total plate count as stated under plate count method.
2. Use MecConkey Agar and VRBGA or VRBLA for Coliform Counts and BGLB broth and indole test for Esch.Coli
3. Use PDA for Yeasts and Moulds
4. Test for Salmonella as describe under Salmonella detection's method. Page 148.

Ice Cream and other Frozen Products:

Since Ice-Creams and other frozen products are made from Milk, dry fruits and number of chemicals, processed in factories, so there are chances of microbiological contamination. It is necessary to assess microbiological quality of the products before distribution and consumption.

SAMPLING: A composite sample is drawn from different packets in a sterile container with a sterile knife after removing the upper layer. When the products are highly frozen the sample is drawn with a Electric drill using sterile kit, thawing is not recommended. In case sample drawn is to be sent to far off places for examination it should be sent in insulated container Minimum lOOgm sample is to be drawn.

CULTURAL EXAMINATION: Melt the sample in a water bath at 50°c for about 15 to 20 minutes. Transfer 10g of melted Ice Cream by a sterilize pipette to a flask containing 90ml of quarter strength ringer solution. Prepare decimal solution up to

10^{-4} or as required. Use these decimal solutions for the following examinations:

1. Use pour plate method for determination of Psychrotrops and mesophiles by using plate count agar or yeast extract milk agar. Incubate at 4-5°c and 20 to 30°c for 14 and 3 days respectively.
2. Use VRBGA or VRBLA for total Enterobacteriacea and Coliform bacteria, incubated at 30°c for 24 hrs respectively. Confirm the presence of E.Coli also as descried under the method of detection of E.Coli.
3. Use PDA or DRBCA for Yeasts and Moulds Count.
4. Use Baird-Parker's medium incubated at 37°c for 24hrs for Staph.aureus.
5. Detection of Salmonella as described under Salmonella see chapter
6. Detection of Listeria monocytogenes - see Page 164
7. Pasteurization test-Thermoduric bacteria count as described under Laboratory Pasteurization test. Page 195

Recommended specifications for Ice Cream:

Aerobic mesophilic	Count (30°c)per g $= m=10^3$ $M=5 \times 10^4$ n=5, c=2
Coliforms (30°c)per g	= m=10, M=100 n=5, c=2
Listeria monocyatogens	absent in 1g
Salmonella	absent in 25g, n=5, c=2

8

Meat and Meat Products Raw and Processed

MEAT AND MEAT PRODUCTS RAW AND PROCESSED

Meat is obtained after slaughtering the animals. There are different processes of slaughtering domestic and commercial processes. In domestic process the carcass is not properly handled and chances of undesirable growth of microorganisms. In slaughter houses carcass is properly handled, rigor is maintained, meat is allowed to soften and tenderized, shifted to cold rooms for chilling, bacterial contamination is controlled. Skin is removed from the carcass and dressed hygienically and cut into pieces to pack.

Cut and exposed surfaces are easily contaminated after dressing and the count increases encromously, the growth in the interior is much less. Counts of aerobes and facultative organisms may developed on the Cut surfaces and anaerobic organisms may develop in the depths of meat.

Aerobic count on the surface of the dressed meat is between 10^3 to 10^5 per square centimeter, out of which few will be psychrotrophic.

Meat on keeping at temperature between 20 to 40°c, Mesophiles will be able to grow rapidly and also chances of anaerobic clostridia to grow. When the Meat is kept below 10°c and surface is moist, Psychrotrophic bacteria such as Pseudomonas will grow and meat surface becomes slimy. On proper aireation the surface layers becomes dried and chances of growth reduces but yeasts and moulds may grow. Mould like Thamniduim, cladospodium (causing black spots) and sportrichum(causing white spot) generally grow in this conditions. It has been seen that when Meat is transported from Chill room packed in Vaccuum packed container to the market for sale the presence of Pseudomonas Acinetobarter microflora have been found.

Microbiological Examination

The surface microflora is assessed by microscopic examination by using contact slides stained by Gram's method. Cultural examination is done by scraping the upper surface by a sterile scraper and collected in a sterile Jar. Allow the meat to thaw if frozen . An alternative method is to cut approximately 50g into small pieces on a sterile tray using sterile knife and forceps.

From above collected samples 10g of sample is transfer to 90ml of peptone water or in a quarter made strength ringer solution. Decimal solutions are made up to 10^{-5} and following examinations may be carried out.

1. General viable counts(aerobic) Use plate count agar and incubate at 4.5°c, 25°c and 37c for 14, 3 and 2 days respectively ISO 2293:1988 specifies 30°c.
2. For Coliforms use VRBA and for 0157:H7 use Trypton sorbital broth with Novobiocin (TSB), Agar(MSA) and confirm by means of E.Coli 0157 latex test kit. On EMB Agar metalic green coloured colonies appear and on MSA pink coloured lactose positive colonies appear.
3. Examination for the presence of Salmonella method is described at page no. 148.
4. For Staphylococcus aureus. Pages 166—169.
5. For viable count an anaerobic bacteria make decimal dilutions with reinforced clostridium medium as a diluent use plate count agar and blood agar, incubate anaerobically at 35-37°c for 2 days.
6. For count of clostridium perfringens use. egg-yolk-free tryptose sulphite iron Citrate cyloserine agar (SICA), for detail method see pages 159—162.

Pork Meat (Sausages, Ham & Beacon, Beef, and others raw Meats

Microflora of these items are common but Sausages, Ham, & Beacon contains often ingridients such as starch, spices and salt which may increase the bacterial load on storage in cold conditions, spoilage organisms. Brochothrix-thermosphacta, Yeast, Lactobacillus and Micrococcus are found. Coliforms organisms, faecal Streptococci may also be found in large numbers. At-higher temperature outside the sausages casing mould Alternaria may grow. Ham gets sour due to the growth of Alcaligenes, Bacillus, Pseudomonas, Lactobacillus, Proteus, Clostridium and others. Refrigerated Packaged products favors the more aerobic bacteria, such as Pseudomonas, Acinetobacter and Moraxella, produce off-flavors, slime and even putrefaction:

Microbiological Examination

A composite sample of 100gm is drawn and homogenized, out of which 10g is transferred to 90ml of quarter strength Ringer's solution. Mix the contents thoroughly on a shaker and prepare decimal solutions in usual way.

Examine as stated under raw meat and in addition following counts:-

1. Lactobacillus count. Using Acetate agar or (Rogosa agar) with cycloheximide(to inhabit microfungi) incubate double-layered pour plates at 30°c for 5 days.
2. Counts of Enterococcus
3. Yeasts and mould counts using dichloran rose bengal chloramphenicol agar, incubated at 25°c for 5 days.

CANNED GOAT, PORK, BEEF AND OTHERS MEAT

Fresh meat after cleaning and deboning or with bones are washed with chlorinated water. Meat is cooled under pressure and mixed with gravy prepared with oil, onions, garlic, spices etc. Cooked meat is also packed in 1 to 2 percent brine. Similarly Pork, Beef and others meats are prepared according to the manufacturer requirement. These products are canned to make them shelf stable and expecting to withstand storage at ambient temperature for at least six to nine months.

In the examination of Canned food, it is rarely possible to obtain a representative sample of the whole since batches may be made up of thousands of cans. It has been recommended by a subcommittee of the international commission for the Microbiological safety of Foods that Random table for drawing sample should be followed. A random sample of 200 cans should be inspected visually for swells and seam defects.

Before microbiological examinations. Cans should be incubated unopened at 30-36° ± 1°C for atleast 14days and at 55°c for 3days respectively. Cans should also be examined before incubation. It is usful to examine the contents of blown cans to determine the organisms responsible for spoilage.

Blown/bulged cans show the presence of thermophilis sporing organisms., Bacillus and Clostridium. This is due to inadequate heat treatment given during processing. Heat sensitive aerobic microorganisms may penetrate through the leaky seam and spoil the products in due course.

MICROBIOLOGICAL EXAMINATION OF CANS:

1. Perform aerobic plate counts on plate count agar at 25°c, 37°c and 55°c.
2. Perform anaerobic plate counts at 37°c.
3. Perform clostridium perfringens, Bacillus cereus count and spores count.

FRESH FISH, FROZEN FISH,DRIED FISH,CURED FISH COOKED FISH, SHELL FISH, PRAWNS, SHRIMPS AND FROZEN PRODUCTS CANNED AND VACUUM PACKED

Spoilage of fish flesh is unique. Fish fat is highly unsaturated and easily oxidized to form free fatty acids which give rancid smell. The spoilage is non-microbiological. Fish flash naturally contains very low levels of carbohydrates, high percentage of protein and non-protein nitrogen fraction differs significantly from that in the meat. Percentage of non-protein nitrogen compound Trimethylamine oxide(TMAO) is higher than other non-protein nitrogen fractions like Urea, creatine etc. TMAO contributes the characteristic odor of the fish. Shell fish such as lobsters have a particularly large pool of nitrogenous extractives ad even more prone to rapid spoilage.

The main source of breeding of these products is water. Bacteriological contamination of these products comes from Water and place of Storage. Water polluted with Salmonella, Vibrios and Coliforms will Contaminate these products. Coagulase-positive staphylococci comes from the persons handling these products.

MICROBIOLOGICAL EXAMINATION

1. Microscopic examination-as explained in meat.
2. Plate culture on Blood agar plate and incubate aerobically and anaerobically.
3. Count of Coagulase-positive staphylococci - use Ba'red - Parker' s medium and incubate at 37°c.
4. Count of Coliforms and E.Coli.
5. Detection of Salmonella
6. Count of Vibrios on TCBS agar, incubate at 36ç° and/or 22°c for 24 to 48 hrs. This is recommended especially for fish from inshore waters.
7. For salt water Fish, Prawns etc use plate count agar plus 15% sodium chloride. Incubate at 4.5°c, 25°c, 37°c for 14, 3 and 2 days respectively.

In this case the diluent should be 1 5% sodium chloride solution.

RECOMMENDED SPECIFICATIONS:

Since microbial spoilage of raw fish, shell fish and Crustacea is very high in short time after the catch, therefore it will always be some risk associated with eating these raw products.

These products of good quality, properly Cooked should have anerobic mesophilic counts less than 500000 per kg, E.Coli less than 10 per g, V.parahaemolyticus less than 10 per g, Salmonella not detected in five of 25g of the samples and staph. aureus not more than 100 in 5 samples.

Poultry and Poultry Products

The initial processing of poultry differs from red meat in a number of ways. During processing birds faeces and feathers contributes to the spread of spoilage microflora as well as pathogens especially Salmonella, Campylobacter, Clostridium perfringens and E.Coli. In the processing plant *substantial* Contamination can occur during defeathering and in the scalding tanks. Refrigerated dressed poultry is spoiled by Psychrotrophic microorganisms, Pseudomonas and Acinetobacter/Moraxella. These mostly grow on the surface. The intestinal tract of poultry contains high number of organisms including pathogens such as Salmonella and Campylobacter.

MICROBIOLOGICAL EXAMINATION

Draw samples as stated in surface sampling techniques. An approximate idea about kind of microorganisms and their relative number is obtained by the use of contact slides or adhesive tape. For quantitative calculations swab procedure is followed. The whole carcass is swabbed or rinsed in a known volume of sterile or rinsed in a known volume of sterile peptone water. Decimal solutions are made and culture examination is carried out as described for raw meat.

EGGS AND EGG.PRODUCTS

General Introduction

Hen and Duck eggs newly laid are generally sterile. Eggshells when comes in contact with faeces, dust, dirty packing material contains number of bacteria. These bacteria include Gram-positive bacteria of the genera Micrococcus, Staphylococcus, Bacillus and Gram-negative bacteria of the general Pseudomonas, Escherichia, proteius, Serratia and Aeromonas.. The infected hens/ducks eggs contents contain Gram-negative organisms and most dangerous is Salmonella enteritidis PT4. When the eggs are broken, there is a chance of contamination from the shell exteriors and on consumption may cause outbreak. Most outbreaks were associated with raw eggs products such as home made MAYONNAISE AND ICE CREAM. Boiled eggs and contents prepared at low heat do not eliminate the microorganisms. Sale of unpasteurized eggs is to be banned.

SHELL EGGS

Microbiological examination of shell eggs contents: Clean the eggs with sterile cloth, dip each egg in alcohol for 10min and allow the spirit to evaporate Mark a hole in the shell with a sterile sharp instrument or using a small carborandum disc on an electric drill. Withdraw contents with a sterile wide-bore pipette. Alternatively crack open the eggs and pour into a sterile jar. Homogenize with a stomacher. Prepare serial dilutions in the usual-way. Examine for arobic mesophilic counts on plate count agar at 25 and 37°c, total Enterobacteriaceae counts, Coliforms counts and presence of Salmonella.

FROZEN WHOLE EGG. Frozen Whole eggs ate generally packed in high density Polymer Pouches or in metal containers with press on lid.

During process contamination chances are there and may contains high counts of bacteria. Bacteria that spoils eggs stored at low temperatures, especially pseudomonos, organisms Alcaligenes, Proteus and Flavobacterium.

MICROBIOLOGICAL EXAMINATION

Draw sample while still in frozen state from the pouch or container after cleaning the pouch with soap solution and chlorinated water, clean metal container and swab with alcohol. keep near the flame and remove the lid, with the help of instrument take out sample from the upper layer and from the centre. Transfer these to a sterilize container. From pouch use a sharp sterile knife, cut from the center and top draw sample with sterile spoon and collect in a container. Allow the frozen samples to soften slightly, immediately blend thoroughly and transfer lOg to 90ml quarter strength Ringer's solution and make decimal solutions as required. Examine the following parameters:-

1 . A diluent microscopic count by the Breed' s smear method.
2. Aerobic mesophilic counts on plate count agar at 25°c and 37°c.
3 Total Enterobacteriaceae using VRBGA Pour plate.
4. Presumptive Coliform Count using VRBLA
5. Detection of Salmonella

The international commission of Microbiological specifications for Foods (ICMF1974) recommended that Frozen whole egg should have a direct microscopic count less than 5×10^6 per gram, a general viable count of less than 10^6 per gram and Salmonella be absent from 25g of sample

WHOLE EGG.POWDER

Eggs are broken, white and yolk of eggs pulp made, dried in similar manner as in milk powder spray drying method and roller dried method. Moisture contents is maintained up to 4 percent.

Egg powder may contain a few hundred microorganisms per gram up to 100 million, depending on the eggs broken and method of handling employed. The same reasons for high numbers hold as for large numbers in frozen eggs. Micrococci, spores of bacteria and moulds will increase on storage.

MICROBIOLOGICAL EXAMINATION

A composite sample is made from the lot and kept in a sterile container. With the help of sterile spoon 10g of sample is transferred to 90 ml of peptone water or quarter strength of ringer solution. Mixing of powder is done by hand or on electric rotor. Carry out all the parameters as described in frozen egg.

10

Vegetable-Salad, Fruit, Dried Fruits-Nuts

Vegetables and Fruits when they come out from the fields remain biologically active for sometime. Their spoilage start when respiration systems become inactive and much depends upon environment in which these are kept. During storage and transporting fresh fruits and vegetables may get spoil after lapse of time.

Fruits are more stable then vegetables because of the fact that fruits contain organic acids(especially citric, malic, tartaric acids) of low pH (usually pH4.5 or lower).These acids acts as preservatives. Fruits also contain high sugar content(especially sucrose, glucose and fructose) which also act as inhibitor of microorganisms. The pH of Vegetables ranges from 5 to 8 Nuts contain high percentage of fats and proteins and low water contents.

Fresh fruits and vegetables may contain normal surface flora and also those from soil and water and perhaps from plants pathogens. Fruits and vegetables grown on land which is fed by organic manures of animal origin or with night soil of human origin contain bacteria such as Salmonella, Shigella,Clostridium botulitinum and psychrotrophic species Listeria monocytogenes. The surface flora generally contain organisms of genera Erivinia and Pseudomonas. Moulds such as Fusarium. Alternaria. Penicillium, Botrytis and Rhizopus may attack on the surface of fruits and vegetables.

The most frequently observed form of spoilage is a softening of the tissue due to the pectinolytic activity of microorganisms.

MICROBIOLOGICAL EXMINATION

The following examinations are performed on the sample of fresh vegetables and fruits.

1. Total viable plate count(2) Pectinolytic organisms may be counted by using polypectate gel medium which is inoculated as spread plates. Incubate the plates at 25-30°c a count is made of the colonies forming depressions in the surface of the gel. The media may be made partially selective for Gram-negative bacteria by incorporating crystal violet and yeasts and moulds may be inhabited by the addition of cycloheximide

When the vegetables in the form of salad are to be consumed and especially kept for later on consumption as in the case of social functions, hotels etc,count of E.Coli must be done. Detection of E.Coli 0157may be done if the vegetables are sourced from the region where the organisms is prevalent. Yeasts and mould count is to be done on storage (vegetables. Shredded Cabbage, raw celery, tamatoes, lettuce and sprouted beans were found as a possible cause of listeriosis in America). In India some cases have been found recently after consumption of Cabbage. Detection of Listeria Monocytogenes is also necessary.

PROPOSED STANDARDS:

Prewashed, fresh vegetables, salad vegetables, beans sprouts and lettuce.

Aerobic mesophilic count	$m=10^5$, $M=10^6$, n=5, c=2
E.Coli	$m=10^2$, $M=10^3$, n=5, c=2
Salmonella	Absent in 25g, n=10, c=0
L.monocytogenes	Not detected in Ig.
Washed frozen vegetables:	
Aerobic mesophilic count	$m=5x10^3$ C = 2, $M=5x10^4$, n=5,
Coliforms	m=10, $M=10^3$, n=5, c=2

FRUIT

The surfaces of fruits contain the natural flora (harmless) and also 'contain contaminants from soil and water. Any injured or spoiled parts will add molds or yeasts. Washing the fruit removes most of the soil microorganisms. Some moulds attack at the time of fruiting and cause disease like black spot or scab. Penicillium italicum and P.digitatum, *Botrytiscinerea* and Monilinia fructigena are common moulds which spoils the fruits.

These moulds can be identified by microscope as stated in the Yeast and Mould chapter.

Prepacked prepared fresh fruit salads or salads kept in fridge for overnight may provide chance for the spread of pathogens. Fruits of high pH (6 to 6.8) like Water Melon and cucumber allow the rapid growth of Salmonella at room temperature.

SPOILAGE OF FRUIT AND VEGETABLE JUICES AND SQUASHES

Juices are prepared by squeezing the fruits or vegetables which have been macerated or crushed. Some time when fruit is brought to the plant it gets spoil in transit or in storage. The juices prepared from these fruits contains lot of *microflora* especially large number of yeasts. Grapes Juice and apples juice generally contain yeasts up to 10^7 per gram. pH ranges from 2:4 to 4.2 of fruit juices and 5 to 5.8 of vegetable juices. Now a days bottled/ canned and pouched juices contain preservatives and pasteurized and as a result lower the amount of *microflora*, Juices are also hot filled in glass bottles and canned at a temperature around 88°c. Hot filled juices may contain Ascospores of Byssochlamys, Talaromyces and Neosartorya may survive, hot filling may produce either mycelial masses or a generalized haze.

Frozen fruit juices concentrates normally show low microbial counts between the range of 10^2 to 10^3 per ml. When these concentrate are reconstituted, packed and stored for some time for marketing, high bacterial counts often develop because of their increased acidity and sugar concentration which favour the growth of yeasts and of acid and sugar tolerant Leuconostic and Lactobacillus species. The change in flavors due to fermentations occurs and become unfit for consumption.

MICROBIOLOGICAL EXAMINATION

1. Microscopic Counts: A known volume of 10^2 dilution is placed on the 1cm^2 slide and mix with one drop of 0.02% erythrosin solution and allow to dry. Count yeast's and mould fragments, sometime bacteria cannot be distinguished because of high solids contents of most fruit juices. Howard mould cell count may be used or a *haemocytometer*.
2. Yeasts and moulds count on DRBCA medium.
3. An aerobic mesophilic count of yeasts, moulds and bacteria on malt extract agar, pH 5.4.
4. An osmophilic count on osmosphilic agar or on orange agar containing 20% sucrose adjusted to pH. 4 or pH 3.5 and incubate at 25-30°c for 7days, examine after 3 and 7 days.
5. Spore Count: Pipette out aseptically 10ml of the sample and transfer to each of two sterile plugged test tubes and into one inset a thermometer through the cotton wool plug so that the thermometer bulb is immersed completely in the sample. Place both tubes in a water bath at 80°c and keep for 15 minutes after the temperature in the control tube just reaches 85°c. Remove the tubes and cool quickly in cold water. Prepare decimal dilutions. Carryout pour plate counts on platecount agar or multiple tube counts using the equivalent liquid medium (glucose tryptone yeast extract broth). Incubate duplicate sets at 30°c and 55°c. for mesophilic spore count and the thermophilic spore count respectively.
6. Ascospore count. Use the procedure described in (5) but heat the tubes for 30 minutes at 70°c in water bath and then use a mould count mediums such as DRBCA or PDA.

CANE, JUICE, SUGAR SYRUPS (SHERBATS) AND SUGARS

These contain high percentage of sugar and acidity(natural and added), favour the growth of yeasts and acid-sugar tolerant Leuconostoc and Lactobacillus species. When sugar syrup is used in the preparation of heat-treated canned fruits, the products should be checked for flat souring organisms-Bacillus stereo thermophilus and gas producing clostridia.

MICROBIOLOGICAL EXAMINATION

Total plate counts for bacteria and yeasts of moulds. For yeasts and moulds use DRBCA, Davis,s yeast salt agar or orange serum agar with 20% sucrose. Incubate at 25-30°c for up to 1 week. Mesophilic and thermophilic spore counts and microscopic examination also be done. In all these products viable counts should not exceed 10 per gram.

HONEY: Today honey is an industrially - marketed food. Artificial hives with moveable inner forms are employed and the honey when ready is removed by uncapping each cell with a knife followed by centrifugal swirling of the frame to draw out the honey without destroying the waxy cellular comb, which can be put back into the hive for use again. Honey rarely contains staphylococci or entric bacteria. It may contains acidophilic and glycolyrtic yeasts which can damage the product. Honey has been found to contain lysozyme, an enzyme with a bacteriastatic as well as lytic effect on most gram-positive bacteria. Suspected honey has been found to contain C.botulinum spores. Honey should be pasteurized at 71°c for 5 min. Honey samples for yeasts and moulds counts and viable spores counts are to be examined.

Cocopowder, Chocolate and Chocolate Confectionery and Sugar Confectionery and Spices

Coco Powder is prepared from coco beans which are the source of pathogenic microorganisms' Chocolate and Chocolate confectionery is prepared from Coco powder and sugar. Most outbreaks has occurred due to consumption of chocolate and Salmonella has been found. Although the temperature during process of Milk Chocolate is maintained between 60 to 75°c but due to low moisture content and dryness Salmonella are protected from heat. Besides this soft fillings of chocolate-covered candies may support the growth of microorganism Chocolate with creamed centers generally contain high coliforms counts and yeasts and moulds as these have been processed at low temperature.

MICROBIOLOGICAL EXAMINATION: Coliform counts, yeasts and moulds count, microscopic examination and detection of Salmonella by using Rappaport-Vassiliads medium as enrichment medium.

PICKLES, FERMENTED VEGETABLE AND PAPADS:

There are two sources of infection of these products. Firstly it come from rotten raw materials used and secondly during processing and storage. Generally pickles and fermented vegetables are not pasteurized before bottled or canned. Preservatives used becomes ineffective after due course of ,time. Microflora of fermented vegetables and papads are different from salted ,Pickled and non-Pasteurized fermented vegetables and usually contain lactic-acid bacteria, particularly Lactobacillus, Leuconostoc and Pediococcus, yeasts and moulds and members of Entrobacteriaceae, Bacillus and Clostridium may cause spoilage of fermented products but some useful lactic and bacteria, yeasts and moulds are used for fermentation and pickling. Pickles preserved in oil and vinegar develop surface growth when they go dry and begin to smell alcoholic.

MICROBIOLOGICAL EXAMINATIONS:

Prepare decimal solutions in appropriate diluents from the mixed gravy, liquor or brine. Fermenting brine or gravy the dilutions may go up to 10^9. The following examination are required.

1. Microscopic examination of the undiluted liquor or brine by using Breed's smear Gram's staining method. For sauces and gravy material dilutions are made. An examination for yeasts and moulds may also be carried out using erythrosin.
2. A viable count of yeasts and moulds and bacteria on malt extract agar at pH 4.5, incubated at 25-30°c for up to 7 days.
3. Yeasts and moulds count by using DRBCA or PDA.
4. Coliform counts using VRBA or by MPN method.
5. A count of lactic acid bacteria on Rogosa agar or acetate agar, prepared as layers plate, incubate at 30°c for 7 days.
6. Counts of Salt-tolerant and halophytic organisms, in case of brine preserved products use 15% Nacl as diluent for these counts and media containing 15% Nacl.
7. Stap-aureus count as usual.

TOMATOE PUREES, TOMATOE SAUCES AND FRUIT PASTES

These products are heat processed and there is little chance of surviving organisms. However some thermoduric organisms, flat sour spores and mould may be present in some products.

Mould Count is done by Howard mould count cell or by *Haemocytometer*. Microscopic examination is done by using erthrosin solution stain, ethanolic solutions of methylene blue or basic fuchin.

Howard mould count on tomato products AOAC method

1. Dilute the product with water and adjust the specific gravity 0.82 at 20°c or refreactometer reading of 45.0-48.7.
2. Add four or five loopsful to the Howard Cell and carefully place the coverslips on the cell,lowering from one side to exclude air bubbles.
3. Adjust microscope to give a field of view with a diameter of 1.382 mm using the × 10 objective. The quantity of liquid in each field of view will then be 0.15 mm^3 when a Howard cell is used. If a microscope is being used that is incapable of such adjustment, determine the field diameter and calculate the required correction factor.
4. Examine at least 25 randomly fields on each of two or more slide preparations, using a mechanical stage to facilitate the choice of fields. The field of view should be selected whilst not looking down the microscope.
5. 'A' positive field is recorded when the aggregate length of not more than three mould hyphae exceeds one sixth of the field diameter. Record the number of positive and negative fields. It is difficult to differentiate fungal hyphae from plant tissue in Howard mould count method. The limit has been set for tomato products by various countries including FDA of USA that maximum 40% positive fields for tomato paste and 20% positive fields for tomato juice.

SPICES AND CONDIMENTS

Spices and Condiments generally are not free from microorganisms though they have been *bacteriostatic* effect on other foods. Different lots of spices vary in their effectiveness, depending on the area of the growth, freshness, storage conditions and whether they have stored as whole or ground up. In India the conversion of raw spices to the table or Kitchen product usually involves only low technology such as sun drying or natural fermentation or pestle mortar grinding. Sun or shade drying is used for leaves like Cinnamon, buds like Cloves and flower stigmas of Saffron, Poppy and Star Anise. Also dried are whole fruits and barriers like the Aniseed, Cardamom, Jeera, Saunf, Pepper and Nutmeg, fruit pulps such as Kokam, Mango and Tamarind, fruits Pods like fenugreek and even the aril of a fruit such as the mace(Javatri). Bulbs and rhizomes of garlic, onion, ginger and turmeric are grown under the soil. These are sun dried. Fermentation of a natural kind to get rid of pulpy material is used in the preparation of Cinnamon bark and Vanilla Pods. The technology of root incision followed by exudation and collection is necessary to obtain Asafetida (Hing). Thus some technology is involved in spice preparation and chances of microbiological contamination at any stage.

The author has been doing microbiology of spices and have seen that some spices carry a heavy load of microflora and bring about contamination of food prepared. It has also been found that small quantity of 5g of curry powder in 500g of curried meat can introduce $5x10^6$ bacteria/g and spoiled the meat. Some spices have been found contaminated with toxin called Aflatoxin $B_1 + B_2$ due to the growth of mould Aspergellus flavous. Ten samples data is tabulated in Table-1. (page 241)

Salmonella was not detected in any sample. It is therefore necessary to examine the spices before these are marketed for consumption.

MICROBIOLOGICAL EXAMINATION

For Whole Spices 50g representative sample after chopping in a sterile pestle and transferred to 450ml of sterile 0.10% peptone water in a flask with glass beads and shaken on a mechanical shaken for 10 to 15 minutes. For spice powder take l0g to 100ml of 0.1% Peptone water. Mix thoroughly by shaking, allow to stand for 1-2 hour.

Serial dilutions are made and examine as under:

1. Total Plate count in usual way, incubate at 37°c
2. Coliforms /E.Coli count in usual way

TABLE-1

Nameof Spices	*Standard plate Count*	*Thermophilic*	*Staphilococcus*	*E.Coli*	*Yeasts & Mould*	*Aerobic spores*
Black-Cardamom	$4x10^5$	$23x10^2$	NIL	NIL	$20x10^1$	NIL
Ginger	$50x10^4$	$20x10^3$	$15x10^3$	NIL	10^3	$20x10^3$
Chillies Pwd	$15x10^5$	$50x10^3$	$10x10^3$	500	100	10^3
Turmeric Pwd	$2x10^6$	$14x10^4$	$4x10^3$	NIL	$2X10^3$	$90x10^2$
Black Pepper	$5x10^5$	$40x10^2$	$60x10^3$	NIL	$50x10^1$	$35x10^2$
Cloves	10x10	20x10	NIL	NIL	NIL	NIL
Cinnamom	10x10	60	NIL	NIL	NIL	NIL
Coriander Pwd	$25x10^5$	$20x10^2$	$21x10^3$	$10x10^2$	$10x10^2$	$20x10^2$
Cumin	$50x10^5$	$60x10^2$	NIL	NIL	$60x10^2$	60
Garlic	$5x10^3$	NIL	NIL	NIL	100	NIL

3. Yeasts and Moulds count in usual way
4. For thermophilies incubate plates at 55°c
5. For staphylococci use staph 110 medium. For confirmation of staph.aureus-see the method under its heading.
6. For spore counts use two diluent tubes 10^{-1}, insert a thermometer into one of the test tubes. Place both the tubes into a stirred water bath set at 81°c, so that the water in the water bath a little above the level of the dilutions in the test tubes. Heat the tubes for I min after the temperature indicatedby the in-tube thermometer reaches 80°c. Remove the tube not fitted with the thermometer and cool rapidly under tape water. Use the heated 10^{-1} dilution to prepare another decimal dilutions series in 0.10% pepfone water. Transfer 1ml of diluent from those tubes to each sterlize plates. Pour plate count agar and mix as usual. Incubate plates at 30°c for 2 days and then determine the aerobic mesophilic spore count.
7. For Clostridium perfringens use sulphite polymxin sulphadiazine agar.
8. For Salmonella detection-see method under its heading Chapter:

INTERNATIONAL COMMISSION OF MICROBIOLOGICAL SPECIFICATIONS FOR FOOD:

(ICMSF)Specifications for Spices:

	n	*c*	*m*	*w*
SPC	5	2	10^4	10^6
MOULDS	5	2	10^2	10^4
E.COLI	5	2	10^2	10^3
SALMONELLA	10	0	0	-

Every European country has her own specifications for spices.

BACTERIAL FREE SPICES

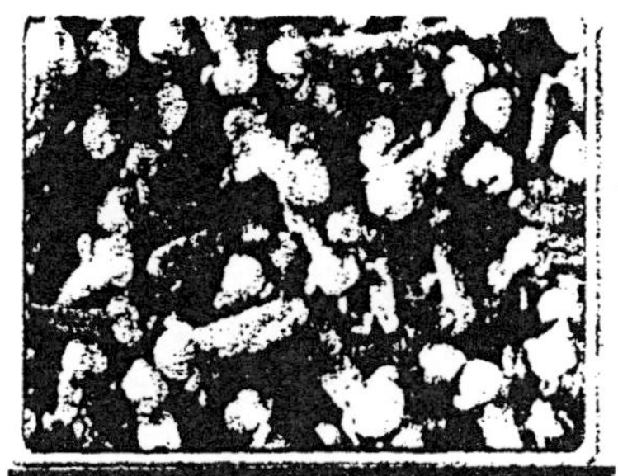

Cloves

Mace

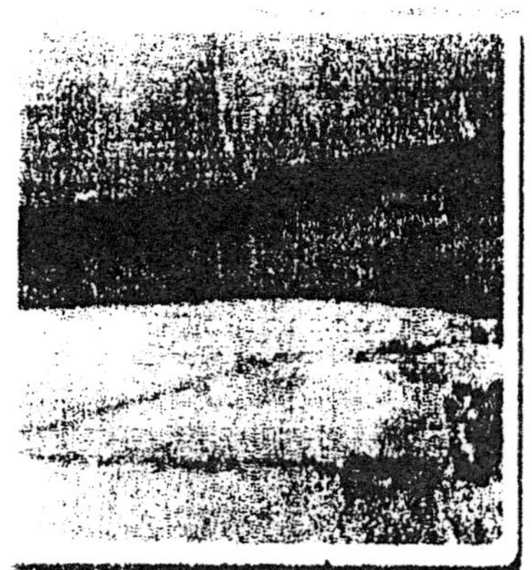

Cassia

Star Anise

12

Soft Drinks and Alcoholic Drinks

Impregination of water with carbon dioxide to give an effervescent liquid and addition of sugar and flavor is called soft drinks. In some soft drinks, fruit juice is added. The source of contamination is either during manufacturing process or from the fruit juices and chances of acid tolerant bacteria such as Lactobacillus.Flavobacterium etc. are always there.

ALCOHOLIC DRINKS

Alcoholic drinks are either prepared from grains or from fruits. These are divided into two categories, Undistilled and distilled liquors, Undistilled liquors such as champagne, Burgundy Sherry etc. are called wines. These are prepared by fermentation of juices with yeasts Saccharomyces cerevisiae or S.elipsoideus or Saccharomyces Carlsbergensis. Yeasts are naturally present on the skin of fruits.

Distilled liquors such as Whisky, Rum, Vodka, Gin, Brandy. Beer etc. are prepared by fermenting grains, cane juice or Molases depending upon the raw material available in a place, the alcohol thus formed is distilled. The distilled alcohol is further matured and blended with esters, colour and essential oils .

Cognac is a term reserved for distilled white Wine made from grapes grown around the French town of that name with a phosphate rich soil.

Microbiological examination. The source of contamination in breweries or distilleries is mash effected by wild yeasts and spoilage bacteria include lactobacilli, pediococci acidic acid bacteria and zymomonas mobilis. Z. mobilis is important spoilage organism in brewing industry, causing turbidity and off flavours.

Chemicals employed in clearing wine may sometime produce sediments. Tin and Copper and their salts have been blamed for cloudiness. Gelatin used in clarifying Wines may cause cloudiness. There have been cases of E.Coli 0157 presence in un-pasteurized Wine.

MICROBIOLOGICAL EXAMINATION

For detection of many wild yeasts use Lysine Agar. S. Cerevisial and S.Carlbergensis are unable to utilize lysine as a sole nitrogen source, whereas many other yeasts can utilize this amino acid. When lysine agar is used to detect Wild yeasts it is important to wash and centrifuge the yeast suspension at least four times using sterile distilled water to ensure that there is no extra source of nitrogen added to the medium. The Pellet is finally resuspended to a known volume using sterile distilled water. Yeast, is Mash effected by wild yeasts, concentrates can be checked for wild yeasts by this method.

For bacterial contamination estimation use plate Count agar with glucose contents increased to 20g. per litre which is added 3g of malt extract per litre, One set of plates being prepared at pH-7 and one set of plates at pH4.0. 1ml of 0-10% stock solution of cycloheximide in 100ml. of the medium may be added before sterilization. Such a medium should provide for the nutrition requirement of most bacteria and yeast likely to be found in these samples. Sets of plates should be incubated aerobically and anaerobically at 30°c.

Distilled liquors containing 40 to 60% alcohol should cause no problems of spoilage by microorganisms.

FLOUR, SPAGHETTI AND MACARONI

Flour is generally incorporated in food products prepared by heat treatment. Most of the microflora includes coliforms which comes from the grains are destroyed during heat treatment. There are chances of organisms like species of Bacillus and clostridium. Bacillus spp. may cause ropiness in bakery products. The use of flour as thickening agent in meat containing products have good chance of Clostridium to multiply.

Spaghetti and Macaroni is prepared from wheat flour. The dough prepared from the flour is converted into these products by extruder and dried. Since there is no heat treatment during dough preparation to final products, both heat-sensitisve and heat-resistant. Microorganisms may be present in the dough and these can grow rapidly during process. Staph aureus and its toxin may cause hazard as subsequent cooking of these products will not inactivate the heat-resistant enterotoxin.

MICROBIOLOGICAL EXAMINATION:

Transfer 10g material to a sterile flask containing 100ml of 0.1% peptone water. Use shaker for adequate mixing, prepare decimal dilutions and carryout the following counts:-

Mesophilic and thermophilic viable counts on plate count agar incubated aerobically and anaerobically at 25, 37 and 55°c for 3 days.

Total Enterobactriaceae or Coliform Count at 30°c using VRBGA or VRBLA.

Aerobic and anaerobic spore counts as stated under spices and condiments page 240.

Staph. aureus- using rabbit plasma fibrinogen agar and enterotoxins as method described under Stap. aureus.

13

Water

When Ice melts, water flows from mountains to land, pick up microorganisms from the soil, number of organisms depends upon the type of surface on which it flows. This water is called natural surface water, flows in the form of stream and stored in lakes and large ponds.

The kinds of bacteria in natural waters are chiefly species of Pseudomonas, chromobacterium, Proteius. Micrococcus, Bacillus, Strephtococcus (enterococci), Enterobacter and Escherichia, Bacteria of the lasts three genera probably are contaminants rather than part of the natural flora.

Contaminated water is a common cause of diseases such as dysentery, cholera, enteric fever and other vibrio diseases, infeactions hepatitis and poliomyelitis gastro enteritis. These diseases are caused by E.Coli, salmonella, shigellae, giardia, enteric viruses and amoebiases.

In food industry the spoilage of food may come from water used as an ingredient, for washing foods for cooling heated foods and for making ice for preserving foods.

E.Coli is used as the most suitable organisms to indicate faecal, Contamination of water; this organism occurs profusely in faecal material from man and animals. Now it is found that many strains originate from the environment and not from faeces. The cysts of protoza, cryptosporidium, parvum and Giardia lamblia in drinking water have been found responsible for very large outbreaks. Concentration methods for detecting crypto sporidium and Giardia have been discussed by UN world Health organization (WHO1993). In UK the method adopted is to collect. Cyst by passing Water through cartridge filter. The trapped material is then eluted, centrifuged, resuspended and subject to flotation in sucrose solution. Giardia cysts are detected microscopically after straining with iodine solution. Cryptosporidium cysts are stained with fluorescent dye-labelled antibody. Now a days improve antibody-based detection systems are available in the market.

Toxins present in Water are produced by Cyanobacteria (blue green algae) such as Anabaensa, Oscillatoria, Aphanizomenon, Nodularia, Microcystis, Noster and Cylindrospermum (WHO 1993 report)

Pseudomonas aeruginosa is another pathogen that is wide spread as a saprophyte in the natural environment. Ps aeruginosa in water, is not an enteritis caused by drinking water, but a contact diseas such as wound infections (WHO 1996)

Aeromonas is also a pathogen which is present in stegnent fresh water and range up to 10^3 per ml (WHO 1996). The water for drinking purposes or for determining potability, it is necessary that the water is free from above stated pathogens. According to India Standards (PFA 1954) and American Public Health Potable Water is treated or chlorinated drinking water.

The absence of normal bacterial index organisms (E.Coli etc.) does not ensure that cryptosporidium, Giardia and Viruses are also absent.

Anaerobic mesophils, E.Coli and Cl. Perfringens counts are useful to know the quality of water.

SAMPLING OF WATER

Glass stoppers wide mouth sterile bottles are used to draw samples of water from taps. When water is chlorinated, the bottle should contain 0.25 ml of 2% solution of sterile sodium thiosulphate for each 250ml of water sample to be collected. The person taking the sample should wear sterile gloves, clean the tap nozzle from outside and inside, let the water should flow for few minutes, turn off the tap, sterilize the tap with alcohol lamp and it cool down by allowing the water to flow. Immediately fill the sample bottle and close with glass stopper.

When source of contamination is to be traced out, sample without sterilization of tap be taken. A swab of the inside and outside of the nozzle to be taken to determine the possibility of contamination of tap itself. Both samples analysis will trace outside the source of contamination.

In collecting water samples from wells, ponds, tubewell,rivers etc. use a glass sterilize mechanical device for holding the bottle and removing the stopper. When the sample is drawn from moving water, the mouth of the bottle should face against the current. In manual collection of a sample, great care is taken so that no water comes incontect with the hands.

MICROBIOLOGICAL EXAMINATIONS

The sample bottle should be shaken 20 to 30 times before opening, remove some water and again shake vertically moving the contents up and down. Prepare aseptically serial decimal dilutions upto 10^{-3} in sterile quarter-strength Ringer's Solution. Examine for the following parameters:

1. Aerobic mesophilic and psychrotroic Counts: Use CPS medium and plate count agar, incubate at 4.5°, 2.5° and 37°c for 14,5 and 2 days respectively. For very low concentration of microorganisms use R2A agar medium.
2 Coliform counts by MPN method. Use three set of tube of each dilutions 10^{2}, 10^{1}, 10^{0} and 10^{-1} quantities of water greater than 1 ml are inoculated into double-strength medium equal in volume to the amount of water inoculated when water sample is suspected of high contamination, dilution higher than 10^{-1} may be tested setting up five to three times tubes of each dilutions.
3. Faecal Streptococci by the multiple tube technique. Use three or five tubes at each of dilutions 10^{2}, 10^{1}, 10^{0} and 10^{-1}. Quantities of water greater than 1 ml are inoculated into double-strength medium equal in volume to the amount of water inoculated.

Incubate the inoculated glucose azide broths at 37°c for 72h.Examine for the production of acid. Record those tubes in which acid is produced, confirm by subculturing a loopful from each positive tube into a fresh single-strength glucose azide broth (5ml per tube) and incubate at 45°c in a water bath for 48h, examining the tubes after 18 and 48h. The production of acid at 45°c within 18h indicates enterococci. If acid is produced after 18h but within 48h it is presumptive evidence of enterococci and can be rapidly determined by microscopic examination for the presence of short-chained streptococci. The most probable number of enterocci can be determined using probability tables (see appendix II)

4. Count for detection of clostridium perfringens

Litmus milk method for gas-producing lactose fermenting clostridia: 200ml of water sample is added to 400ml of sterile litmus milk (heated to 100°c to drive off dissolved air and in a flask of 1000ml capacity. Heat in a water bath to 80°c for 10min.Cool, Cover the surface of the milk with melted sterile 'vaspar' and incubate at 37°c for 5 days.

Examine every 24h for the production of acidity, cloting and gas—the reaction known as '*stormy-clot*-which indicates the presence of cl.perfringens. Generally the number of cl. Perfringens present are very low so the quantity of water sample for testing should not be less than 100ml

MPN method: Use DRCM as the selective medium and confirmation by streaking on plates of willis and Hobb's lactose egg-yolk milk agar and blood agar plates. See method under C.Perfringens.

Wilson and Blair's sulphite medium for sulphite reducing clostridia:

Add 20ml of water sample(previously heated to 80°c for 10 min to destroy vegetative forms of bacteria) to 20ml of Wilson and Blair's sulphite medium(dispensed in longneck tube) mix well and allow to set.

Incubate aerobically at 37°c for 24h. After incubation see the presence of black colonies(sulphite reducing) and confirm for cl. Perfringens.

TEST FOR PSEUDOMONAS AERUGINOSA COUNT

Prepare plates of cephaloridine. Fucidin-Cetrimide agar(CFCA) and inoculate with 0.lml of dilutions. Spread with sterile bend glass rod. Incubate first 30°c for 4h and then at 41+1°c for 44h. Colonies showing yellow-green fluorescence are assumed to be Pseudomonas aeruginosa. Confirm its presence with Hugh and Leifson;s medium. For smaller number membrane filtration method is used.

TEST FOR AEROMONAS

Use spread plate method for inoculation or membrane filtration to detect smaller numbers Use modified XLD Dagar for growth. Water sample drawn from house old tap or from main distribution systems incorporate EDTA sodium salt at 50mg per litre to protect the organism against any traces of copper.

MEMBRANE FILTRATION METHODS

These methods are very useful when small number of organisms are suspected in water or any liquid foods.

In water sample Aerobic mesophilic counts can be carried out by using membrane type nutrient broth or membrane type tryptone soya broth and incubating the filters at 37°c for 18-24h or at 25°c for 2days. For counts of E.Coli use Millipore type type HC filters. Incubate the filters for 2h at 37°c on pads saturated with membrane type nutrient broth and then transfer the filters to pads saturated with membranetype. MacConkey's broth and incubate for 18h at 44°c. In this way E.Coli can be detected within 2 working days. Enterococcus faecalis counts should be determinted with preselective incubation of filters on nutrient broth for 2h at 37°c before transferring the filters to pads saturated with glucose azide broth and incubating at 45°c for 18h.

MOLECULAR METHODS FOR MICROBIOLOGICAL DETECTION

In classic culture technique one especially difficult problem in the analysis of drinking water is that small numbers of organisms have to be detected in large volumes of Water, secondly time consuming, elaborates and expensive.

Now there is a growing interest in using molecular techniques in microbial analysis of drinking water. For example, molecular techniques can detect fragments of nucleic acids that contain a specific sequence indicative of a certain pathogen. Even if only one cell (and so only one single fragment) with the desired nucleic acid sequence is present, then PCR-method (polymerase chain reaction) can in theory, selectively multiply this fragment until a detectable number of fragments has been produced. Other method use antibodies that in, they are based on the immunological detection of cell components specific for the pathogens of interest. The coupling of antibodies or nucleic acid fragments to dyes can further simplify their detection. These methods have the advantages for the detection of viruses as well.

MEDIA

A selection of media commonly used in food microbiology is given in this book.

Apart from microbiology examination of food by direct methods, there is full microbiological examination method which require a media to encourage the growth in liquid form, or in solid form. Agar is one of the ingredients of these medias. It is a poly-saccharine with a remarkable properties, which is produced by species of red Alae

It is a complelx and variable material, the major component of agar is agarose which is made of alternating units of 1, 4 linked 3, 6 anhydro-L-galactose(or L-galactose) and 1-3, linked D-galactose (or 6-0 methyl -D-galactose). It forms a gel at low concentration (1.5-2%) and melts at the temperature of boiling water and solidifies when cooled to about 40°c. It has no effect on bacteria and stable to microbial hydrolysis. Only a relatively small group of micro-organism are able to degrade agar, presumably due to the presence of the unusual L-form of galactose in the polymer. It does not provide any nutrient to bacteria, Numerous materials are added in different culture media, carbohydrate, serum, whole blood, bile, ascitic fluid and hydrocele fluid. Carbohydrates are added to increase the nutritive value of the medium and to indicate the fermentation reaction of the microbes being studied, serum, blood and ascetic fluid are added to promote the growth of fastidious organisms. Dyes are added to the culture media to act as indicators to detect the formation of acid when fermentation reactions occurs or act as inhibitors of growth of certain bacteria but not to others. An example of an indicator dye is phenol red which is red in an alkaline or netural medium but yellow in an acid one. An example of an inhibitory dye is gentian violet which inhibits the growth of most Gram + ve bacteria.

Most of the media described here are available in either dehydrated or ready prepared form from many manufacturers, powder or in granular form which may be reconstituted in distilled water. Care should be taken to avoid inhaling or ingesting dust while dissolving in Water, wear a face mask. Many ingredients of selective media, including many antibiotics, are toxin or irritant.

ACETATE AGAR

A selective medium for the isolation or enumeration of lactobacilli.

1. Basal Medium

Peptone	10.0g
Meat extract	8.0g
Yeast extract	5.0g
Agar	15.0g
Triammonium citrate	2.0g
Salt Solution	5.0ml
D Glucose	10.0g
Tween 80	0.5ml
Distilled Water	1 litre
pH	5.4

SALT SOLUTION

Magnisum Sulphate. Hydrated (Mag SO4.7.H2 0)	8.0g
Manganese Sulphate, hydrated(Mn SO4.4H2 0)	2.0g
Distilled Water	100ml

Dissolve the peptone, meat extract, yeast extract and agar in 1 litre of distilled water by autodaving at 121°c for 15 min. Then add the citrate and salt solution and mix well. Adjust the pH to

5.4 and filter. If necessary add the glucose and Tween 80 and mix thoroughly. Dispense in 90ml amounts in screw-capped bottles.sterilize by autoclaving at 121°c for 15 min.

Preparation of 2M acetic acid-sodium acetate buffer at pH 5.4

Dissolve 23.3 g of sodium acetate(Hydrated) and 1.7g of glacial acetic acid in distilled water and Make up to 100ml. Check the pH by diluting a portion of 1:10 in distilled water before using the pH meter (dilution is necessary because the high concentration of sodium ions may effect the glass electrode). Distilled the screw-capped containers in 10ml amounts and sterilize by autoclaving at 115°c for 20min.

Preparation of the medium for use:- The medium may be used with or without the acetic acid-sodium acetate buffer (when the products contains high percentage of salt, buffer is not required) When the medium is to be used with buffer added, add aseptically 10ml of the buffer solution to 90ml of the medium, which has been melted and cooled to 50°c. Mix gently by careful inversion to avoid frothing and pour double layer plates in the usual way. The final pH of the complete medium should be 5.4 ± 0.05.

An important media's used in food microbiology are:

Medium:	*Use*
1. Baired-Parker agar isolation	A selective and diagnostic medium for and enumeration of staphylococcus aureus.
2. Brillian green Lactose Broth	MPN of Coliforms in food and also for isolating and counting.
3. Buffered Peptone water	For non-selective resuscitation.
4. Crystal violet/Azide blood agar	For Enumeration of faecal Streptococci
5. Cefixime/Tellurite-Sorbitol Mac Conkey Agar	For the isolation and differentiation of Escherichia Coli 0157.
6. CPS Medium	For the isolation and Enumeration of aquatic bacteria
7. Dichloram 18% gly cerol Agar(DG-l 8)	For the enumeration and isolation of xerophilic fungi in foods

Medium:	*Use*
8. Dichloram-Rose Bengal agar(DRBCA)	chloramphenicol Enumeration of yeasts and Moulds.
9. Differential Reinforced Clostridial Medium	For the detection and enumeration of sulphite reducing clostridia by MPN method.
10.Egg-yolk Agar/Broth	For the detection of lecithinase production.
11.Fortified Nutrient Agar	For spore production by Bacillus stearothermophilas.
12.Giolitti and Contoni's Tellurite Mannitol glycine Broth	For selective enrichment of Staphylococcus aureus
13. Glucose Azide Broth	For enumeration of faecal streptococci by MPN method.
14. Hektoen Enteric Agar	For selective isolation of Salmonella and Shigella.
15.Hugh and Leifson's Medium modified	For differentiating oxidative and fermentative metabolism of Glucose and mannitol by *Staphylococcus* and *Micrococcus*.
16.LitmusMilk	For Milk spoilage organisms
17. MacConkey's agar	For isolation and growth of enterobacteria. Especially the Coliform bacteria.
18. MacConkey's Broth	For the detection and enumeration of lactose-fermenting enterobacteria by MPN method.
19. Malt Extract Agar	For the culture of yeasts and moulds.
20. Mannitol Egg-yolk Phenol Red Polymyxin agar	For detection and differentiation of Bacillus Cereus.
21. Modified Alkaline Peptone Water	For the isolation of Vibrio.
22.Nutrient Agar and Nutrient Broth	A general purpose culture medium for bacteria.
23 Rappaport-Vassi liaidis Broth	Selective enrichment of Salmonella.
24. Selenite cystine broth	Selective enrichment of Salmonella.

Medium:	*Use*
25. Peptone Water Diluent	For enumeration and isolation techniques
26. Plate Count Agar (Tryptone glucose Yeast Extract Agar)	For General viable counts of bacteria.
27. Polymyxin Pyruvate Egg-yolk Mannitol Bromthymol Blue Agar (PPE MBA)	For the enumeration of BacilliusCereus in foods.
28. Potato Dextrose Agar	For the growth of microfungi.
29. Ringer:s solution quarter strength	Used as diluent and suspending liquid.
30. Rogosa Agar Modified	For the isolation and enumeration of lacto bacilli in fermented milk.
31. Salmonella-Shigellae Agar(SSA)	For the isolation of Salmonella and shigellae
32. Salt Polymyxin Broth	For the enchricment of Vibrio.
33. Sorbital MacConkey;s Agar (SMAC)	For the isolation and differention of Escherichi coli 0157.
34. Taylor's xylose Desoxycholate (XLD)Agar	For selective isoltion of Salmonella.
35.Triple Sugar Iron Agar	For differentiation of members of *Enterobacteriaceae*
36. Thiosulphate citrate Bile-salts sucrose Agar (TCBS)	For isolation of Vibrio
37. Violet Red Bile glucose Agar	For the enumeration of the Entero-bacteriaceae.
38.Violet Red Blue Lactose Agar enterobacteria	For the enumeration of lactosefermenting
39. Wilson and Blair's sulphite Medium	For Clostridium enumeration
40. Yeast Extract Milk Agar	For detecting bacteria in milk and milk products.

SHIGELLAE

The genus Shigella consists of Gram-negative, aerobic and faculatively anaerobic, non-sporulating, non-motile rod-shaped bacteria in the family Enterobacteriaceae. They do not decarboxylate lysine and gas production is restricted to one serotype (S.flexneri 6). Lactose is usually not utilized and when there is fermentation it is usually delayed for several days. The Shigella Consists of four species: Shigella dysenterial (Serogroup A) and S.flexneri (Serogroup B) Shigella are even more difficult to isolate from foods than from clinical specimens, their survival is influenced by temperature, pH, and types of food. They survive longest when holding temperatures are 25°c or lower. S.flexneri and S.Sonnei survive for over 170 days in flour and milk, but for less time in high-acid foods.

METHOD-PLATE COUNT METHOD

To 25g of sample in a stomacher '400' filter bag, 225ml Gram-negative broth is added and homogenized for 1min. in Stomacher blender(Colworth 400), the homogenate was enriched for 16-12h at 37°c. MaConkey's and desoxycholate citrate (DCA) and xylose-lysine, desoxycholate XLD agar plates were streaked from the homogenate before and after enrichment. Five typical Colonies from each selective plates were streaked for purity on nutrient agar. Oxidase-negative strains were picked from the NA plates, streaked on the slant and stabbed into the butt of triple sugar, gave typical colonies. Shigella reactions were confirmed biochemically With the API 20E system(bioMerieux) and serologically with group-specific somatic antiserum(available from Biological Production, Izat Nagar Bareilly) (UP) India or Swedish Institute for infectious diseases control, Solna, Sweden)

Detection of Shigella Spp in food with nested Polymerase Chain reaction (PCR)

In some advance Microbiological laboratories PCR methods have been developed but a universal cultural media for PCR for all species has not been developed so far.

Presently in Sweden National Food administration up sale Laboratory a nested PCR method has developed. The presence of PCR inhibitors in food samples may inhibit amplification and limit the usefulness of PCR techniques. The buoyant density centrifugation method is successful in food as separation of bacteria and food due to their different buoyant densities in gradient medium. PCR method can be completed within two days of working.

General Media

BAIRD-PARKER'S MEDIUM (EGG-YOLK TELLURITE GLYCINE AGAR)

(A selective and diagnostic medium for isolation and enumeration of *Staphylococcus aureus)*

Tryptone	10.0g
Lab-Lemco meat extract	5.0 g
Yeast extract	1.0g
Lithium chloride, hydrated	5.0g
Agar	20.0 g
Sulphadimidine sodium salt (0.2% solution)	25.0ml
Distilled water	975 ml

pH 6.8-7.0

Dissolve the ingredients by steaming. Adjust to pH 6.8. Dispense without filtration in 90-ml amounts in screw-capped bottles and sterilize by autoclaving a 121°C for 15 min. Before pouring the plates, to each 90ml of basal medium, melted and cooled to 50°C, add aseptically the following solutions (all previously sterilized by filtration) in the amounts given:

(a) 20% (w/v) solution of glycine	6.5 ml
(b) 1% (w/v) solution of potassium tellurite[(H)] (e.g. from Sigma or Merck)	1.1 ml
(c) Egg-yolk emulsion	5.4ml

Mix well and pour into Petri dishes in 12-ml amounts. These poured plates may be stored in plastic bags at 4°C for up to 1 month. Before use 0.5 ml of a 20% (w/v) solution of sodium pyruvate should be spread aseptically over the surface of each plate. The plates are then dried (with the medium *surface* uppermost) at 50' C for 1 h, before inoculation.

Note

1. The 0.2% solution of sulphadimidine sodium salt is prepared by dissolving 0.5 g of pure sulphadimidine (sulphamezathine) in 25ml of 0.1 N sodium hydroxide solution and making up to 250 ml with distilled water.
2. The source of potassium tellurite may affect the selectivity and sensitivity of this medium. It is possible to obtain standardized sterile potassium tellurite solution or egg-yolk tellurite mixture from manufacturers of dehydrated media. However, if the potassium tellurite solution is prepared in the laboratory, each bottle of potassium tellurite from any batch not previously tested should be validated by checking the suitability of the medium against known stock cultures of *Staphylococcus aureus*.
3. The method of preparation of the egg-yolk emulsion is given para 412. It must be stored in a refrigerator.
4. The sodium pyruvate solution must be stored in a refrigerator and used within 1 month.
5. As indicated, this medium is available in dehydrated form, as a basal medium, to which must be added egg-yolk, tellurite and the sulphadimidine salt. A ready-prepared egg-yolk tellurite emulsion is available from, for example, Oxoid or Difco, and can be used in conjunction with these dehydrated media.

(Egg-yolk emulsion—seperate the yolk from the white of eggs by pipette and add four parts of distilled water to one part of egg yolk. Mix thoroughly and heat in a water bath at 45° for 2 h. centrifuge to remove the precipitate. Decant the supernatant liquid. Sterilize by fitration through stached graded membrane filters)

BRILLIANT GREEN LACTOSE BILE BROTH

(Selective medium for isolating and counting coliform organisms and *Escherichia coli)*

Peptone	10.0g
Lactose	10.0g
Ox bile (Oxoid L50 or equivalent)§	5 20.0 g
Brilliant green (1% aqueous solution)	13.3ml
Distilled water	to 1 litre

pH7.4

§ Different proprietary preparations may not have equivalent activity at the same concentration.

Dissolve the peptone and lactose in 500 ml of distilled water. Dissolve the ox bile in 200 ml of distilled water. Mix the two solutions, add distilled water to 950 ml, and adjust to pH7.4. Add 13.3 ml of a 1% aqueous solution of brilliant green and then make up the volume to 1 litre with distilled water. Dispense into test-tubes containing inverted Durham tubes, plug and sterilize by autoclaving at 121°C for 15min.

(Mackenzie *et al.*, 1948)

BUFFERED PEPTONE WATER

(For non-selective resuscitation)

Peptone	10.0 g
Sodium chloride	5.0g
Disodium hydrogen phosphate	3.5g
Potassium dihydrogen phosphate	1.5 g
Distilled water	to 1 litre

pH7.2±0.2

Dissolve the ingredients in the water, distribute as required, and sterilize by autoclaving at 121°Cfor 15 min.

CPS MEDIUM

(for the isolation and enumeration of aquatic bacteria)

Soluble casein (Merck)	0.5 g
Peptone	0.5 g
Soluble starch	0.5 g
Dipotassium hydrogen phosphate	0.2 g
Magnesium sulphate, hydrated (MgSO4-7H2O)	0.05 g
Ferric chloride (0.01% solution)	4 drops
Glycerol	1 ml
Agar	15.0 g
Distilled water	1 litre

pH 6.9-7.0

Dissolve the casein, peptone, starch and agar in the water by steaming. Add the remaining ingredients to the hot medium, mix well and filter. Distribute as required and sterilize by autoclaving at 121°C for 20min.

(Collins *et al.*, 1973)

CRYSTAL VIOLET AGAR

(A medium for the isolation of Gram-negative bacteria)

Yeast extract agar, plate count agar or nutrient agar	100 ml
Crystal violet (0.05% aqueous solution)	0.4ml

Add aseptically the crystal violet solution (previously sterilized by filtration) to the sterile molten medium immediately before pouring the plates. This gives a final concentration of crystal violet of 2 p.p.m.

(Holding, 1960)

DICHLORAN-ROSE BENGAL-CHLORAMPHENICOL AGAR (DRBCA)

(For the detection and isolation of fungi in foods)

D-Glucose	10.0g
Mycological peptone	5.0g
Potassium dihydrogen phosphate	1.0g
Magnesium sulphate, hydrated (MgSO4-7H2O)	0.5g
Rose bengal, 5% (w/v) aqueous solution	0.5 ml
Dichloran, 0.2% (w/v) ethanolic solution	1.0 ml
Chloramphenicol, 1% (w/v) ethanolic solution	10.0ml
Agar	15.0g
Distilled water	1 litre

Steam to dissolve. Dispense and sterilize by autoclaving at 121 °C for 15min. Dispense in 15-ml amounts into sterile Petri dishes.

Use immediately, or store refrigerated in the dark. This medium, when made, *must* be kept in the dark, because under the influence of light Rose Bengal may generate cytotoxic products which will result in the medium becoming too inhibitory (Banks *et al.*, 1985)

(Pirt and Hocking, 1985)

DICHLORAN 18% GLYCEROL AGAR (DG18)

(For the enumeration and isolation of xerophilic fungi in foods)

D-Glucose	10.0g
Mycological peptone	5.0 g
Potassium dihydrogen phosphate	1.0 g
Magnesium sulphate, hydrated ($MgSO_4.7H_2O$)	0.5g
Dichloran, 0.2% (w/v) ethanolic solution	1.0ml
Chloramphenicol, 1% (w/v) ethanolic solution	10.0ml
Agar	15.0g
Distilled water to 1 litre	
Glycerol 220 ml	

Suspend the ingredients except the dichloran, chloramphenicol and glycerol in about 800ml of water. Heat to dissolve. Add the ethanolic solutions of dichloran and chloramphenicol to the mixture, and make up to 1 litre with water (at 55°C). Add the glycerol (preheated to 55°C), mix thoroughly, and sterilize by autoclaving at 121°C for 15min. Cool to 50°C, mix well, and dispense into sterile Petri dishes. Allow to set, and dry.

Use immediately, or can be stored for up to 7 days in the refrigerator.

(Hocking and Pitt, 1980)

DIFFERENTIAL REINFORCED CLOSTRIDIAL MEDIUM

(For the detection and enumeration of sulphite-reducing clostridia by a multiple tube technique)

Basal medium

Peptone	10.0g
Lab-Lemco meat extract	10.0g
Sodium acetate, hydrated	5.0 g
Yeast extract	1.5 g
Soluble starch	1.0g
D-Glucose	1.0 g
L-Cysteine monohydrochloride	0.5 g
Distilled water	1 litre

pH 7.1-7.2

Dissolve the peptone, meat extract, sodium acetate and yeast extract in 800ml of distilled water. Mix the soluble starch to a paste with a little of the remaining 200ml of water, boil the rest of the 200 ml of water and stir it into the paste. Mix the two solutions and steam for 30 min to complete the solution of the ingredients. After the steaming add the glucose and cysteine, mix and adjust the pH to 7.1-7.2. Filter through hot paper pulp (see p. 445), dispense in 25-ml amounts in 30-ml (1 oz) screw-capped (McCartney) bottles. Sterilize by autoclaving at 121°C for 15min.

Sodium sulphite solution

Prepare a 4% (w/v) solution of sodium sulphite (anhydrous) and sterilize by filtration. This solution may be stored in fully filled screw-capped bottles in the refrigerator for up to 14 days.

Ferric citrate solution

Prepare a 7% (w/v) solution of ferric citrate (scales), heating briefly to dissolve. Sterilize by filtration. It may be stored in fully filled screw-capped bottles in the refrigerator for up to 14 days.

Preparation of complete medium

When the DRCM is required, steam and cool the basal medium, and then add aseptically to each bottle of medium 0.5 ml of a freshly prepared mixture of equal volumes of sodium sulphite solution and ferric citrate solution.

The medium may be made more selective for sulphite-reducing *Clostridium* spp. by also adding aseptically to each bottle 0.9ml of a solution of 10000 units of polymyxin B sulphate in 5 ml of sterile distilled water.

(Gibbs and Freame, 1965; Freame and Fitzpatrick, 1971)

EGG-YOLK AGAR

(For the detection of lecithinase production)

Egg-yolk emulsion	10ml
Sodium chloride	1.0 g
Yeast extract agar or nutrient agar	100 ml

To the basal medium add the extra salt and sterilize by autoclaving at 121°C for 20min. Cool to 45°C and add aseptically the sterile egg-yolk emulsion, mix well and pour plates. The egg-yolk emulsion (see p. 262) may be prepared after the method of Billing and Luckhurst (1957). Alternatively, egg-yolk emulsions are available commercially from most suppliers of dehydrated media and media supplements.

FORTIFIED NUTRIENT AGAR

(For spore production by *Bacillus stearothermophilus*)

Nutrient agar powder	15 g
Plain agar powder	5 g
D-Glucose	0.5 g
Manganese sulphate, hydrated (MnSO4-4H2O)	30 mg
Distilled water	1 litre

Dissolve the ingredients in the water by steaming. Dispense in Roux bottles or in 500-ml screw-capped medical flat bottles in sufficient amount to provide an approximately 2-cm layer when the bottles are laid on their sides. Sterilize by autoclaving at 121°C for 25 min, and place the bottles on their sides so that the medium sets to provide the maximum surface area.

(Finley and Fields, 1962)

GIOLITTI AND CANTONI'S TELLURITE MANNITOL GLYCINE BROTH

(For the selective enrichment of *Staphylococcus aureus*)

Tryptone	10.0g
Meat extract	5.0g
Yeast extract	5.0g
Lithium chloride	5.0g
Mannitol	20.0 g
Sodium chloride	5.0g
Glycine	1.2 g
Sodium pyruvate	3.0 g
Distilled water	1 litre

pH 6.9 ±0.2

Dissolve the ingredients in the water with heating, mix well. Cool to 25°C and adjust the pH. Distribute as required, and sterilize at 115°C for 20min.

The sterile basal medium may be stored refrigerated for up to 15 days.

Potassium tellurite solution

Potassium tellurite[H]	1.0 or 10.0 g*
Distilled water	to 100ml

Dissolve, mix and sterilize by membrane filtration. The quality of potassium tellurite is critical. Potassium tellurite from Merck has been found satisfactory; its inhibitory action for known strains of *Staph. aureus* should be determined, especially if other sources for the tellurite need to be used.

* If the enrichment is to be achieved from adding a 10^{-1} dilution, then the lower concentration of potassium tellurite is used; if the enrichment broth is to be inoculated with undiluted food (up to 1 g of food per 20ml of broth or 5 g food per 100ml of broth) then the higher concentration of potassium tellurite should be used to suppress the higher numbers of contaminants that will find themselves in a medium more substantially modified by the added food.

Preparation of complete medium

To the sterile basal medium add with aseptic precautions an appropriate volume to an equivalent of 0.5ml of sterile potassium tellurite solution per 100ml of basal medium (e.g. 0.1 ml per 20 ml). The choice of concentration of potassium tellurite solution is determined by the type and amount of inoculum (see above). Use the complete medium on the day of preparation.

(Giolitti and Cantoni, 1966)

GLUCOSE AZIDE BROTH

(For the enumeration of faecal streptococci by the multiple tube technique)

Peptone	10.0g
Sodium chloride	5.0 g
Dipotassium hydrogen phosphate	5.0g
Potassium dihydrogen phosphate	2.0 g
D-Glucose	5.0 g
Yeast extract	3.0g
Sodium azide(H)	0.25 g
Bromcresol purple (1.0% solution)	3ml
Distilled water	1 litre

pH 6.6-6.8

Dissolve the ingredients in the water. Distribute in 5-ml amounts in 150 x 16-mm test-tubes. Sterilize by autoclaving at 121°C for 15min. For inocula of large amounts of sample or diluent (5ml or more), a double-strength medium should be prepared and distributed in amounts equal in volume to the inocula to be used. Double-strength medium is prepared in a similar manner to that described above, the ingredients being dissolved in half the quantity of distilled water.

Note: The effectiveness of this medium should be checked from time to time using stock cultures at low inoculum levels; it has been found that variations in the nutrients composition (particularly the peptone) from batch to batch may affect selectivity and/or sensitivity.

(Hannay and Norton, 1947)

HEKTOEN ENTERIC AGAR

(For the selective isolation of *Salmonella* and *Shigella)*

Proteose peptone	12.0 g
Meat extract	3.0g
Lactose	12.0g
Sucrose	12.0g
Salicin	2.0 g
Sodium chloride	5.0g
Bile salts§	9.0 g
Sodium thiosulphate	5.0g

Ammonium iron (III) citrate*	1.5 g
Acid fiichsin	0.1 g
Bromthymol blue (0.4 % aqueous solution)	16 ml
Andrades' indicator	20ml
Agar	14.0g
Distilled water	1 litre

pH 7.5 ±0.2

§ Different proprietary preparations may not have eqüivalent activity at the same concentration.

* The ammonium iron(III) citrate used should be the more pure green form (around 18% iron content).

Mix the ingrediénts in the water. Allow to soak for 15 min, with occasional mixing. Heat to boiling for a few seconds until the agar is dissolved. Do not autoclave. Cool to 55°C and pour into Petri dishes.

Citrobacter and *Proteus* may be further inhibited by incorporating novobiocin in the medium just before pouring the plates (Hoben *et al*, 1973). To each 1 litre of medium add 1ml of a sterile novobiocin solution (150 mg novobiocin dissolved in 10ml of distilled water, sterilized by membrane filtration).

(after King and Metzger, 1968)

HUGH AND LEIFSON'S MEDIUM

(For differentiating oxidative and fermentative metabolism of carbohydrates, by Gram-negative bacteria)

Recipe 1 (Hugh and Leifson, 1953)

Peptone	2.0 g
Sodium chloride	5.0g
Dipotassium hydrogen phosphate	0.3 g
Bromthymol blue (1% aqueous solution)	3.0 ml
Agar	3.0 g
Distilled water	1 litre

pH 7.1 ±0.2

Recipe 2 (Scholefield, 1964)

Tryptone	1.0g
Yeast extract	1.0g
Sodium chloride	5.0g
Dipotassium hydrogen phosphate	0.3 g
Bromthymol blue (1% aqueous solution)	3.0ml
Acid fuchsin (1% solution)	1.5ml
Agar (Oxoid No. 3)	4.5 g
Distilled water	1 litre

pH 7.0 ±0.2

Add the ingredients, except indicators, to the water and dissolve by steaming. Add the indicator(s) and mix well. Adjust the pH to 6.8 with thorough mixing, using a pH meter. (After autoclaving, the final pH of the medium should be 7.0-7.1.) Dispense in 10-ml amounts in 150x 16-mm test-tubes, and close with Astell seals (Astell-Hearson), or other suitable tube closures. Leave the closures loose, and sterilize by autoclaving at 121 °C for 15min. To each tube of molten medium add aseptically 1 ml of a sterile 10% solution of the desired substrate solution, mix well (without aeration) and allow to set. The carbohydrate normally employed in this medium is glucose, and it may be incorporated into the medium at the time of preparation if differential studies of reactions with a range of substrates are not being undertaken.

The second recipe has been found to allow the use of a single tube only, provided the tubes of media are stood in a boiling water-bath for 10min and then cooled rapidly immediately before inoculation; in addition, a very clear colour change from blue-green to orange-red results from the production of acid.

LITMUS MILK

Add sufficient litmus solution to reconstituted skim milk to give a pale mauve colour (10ml of 4% litmus solution per litre of milk). Dispense as required (in relatively small volumes) and sterilize at 121°C for 5min, follozwed by steaming for 30min on each of the two following days. Autoclaving on the first day has been found to reduce substantially the number of spoiled tubes or bottles resulting from residual spores. Always preincubate before use.

MAcCONKEY'S AGAR

(For isolation and growth of enterobacteria, especially the coliform bacteria)

Peptone	20.0 g
Bile salts§	5.0 g
Sodium chloride	5.0 g
Lactose	10.0g
Neutral red (1% aqueous solution)	7.0ml
Agar	15.0g
Distilled water	1 litre

§ Different proprietary preparations may not have equivalent activity at the same concentration.

Dissolve the peptone, bile salts and sodium chloride in the water by steaming. Cool and adjust to pH 7.4. Add the agar and dissolve by autoclaving. Filter through hot paper pulp. Adjust the pH to 7.4. Add the lactose and neutral red and steam until dissolved. Mix well and distribute in bottles or test-tubes as required. Sterilize at 115°C for 15min.

(Report, 1969)

Brilliant green MacConkey's agar

Add 3.3 ml of 1% brilliant green solution at the same time as the lactose and neutral red (Harvey and Price, 1974).

MAcCONKEY'S BROTH

(For the detection and enumeration of lactose-fermenting enterobacteria by the multiple tube technique)

Peptone	20.0 g
Bile salts§	5.0 g
Sodium chloride	5.0 g

Lactose	10.0g
Bromcresol purple (1% solution)	1 ml
Distilled water	1 litre

pH7.4

§ Different proprietary preparations may not have equivalent activity at the same concentration.

Add to the water the peptone, bile salts and sodium chloride and heat in a steamer for 1-2h. Add the lactose and dissolve by heating for a further 15 min. Cool and filter. Adjust to pH7.4. Add the indicator. Mix well. Distribute in 5-ml amounts in 150 x 16-mm test-tubes provided with inverted Durham tubes. Sterilize by autoclaving 115°C for 15 min. Double-strength medium can be prepared in a similar manner, by dissolving the ingredients in half the quantity of distilled water. Distribute in amounts equal in volume to the inocula to be added.

(Report, 1969)

MAcCONKEY'S BROTH (FOR MEMBRANE FILTRATION)

(For the detection and enumeration of lactose-fermenting enterobacteria by membrane filtration)

Peptone	10.0g
Bile salts§	4.0 g
Sodium chloride	5.0g
Lactose	30.0 g
Bromcresol purple (1% solution)	12ml
Distilled water	1 litre

pH7.4

§ Different proprietary preparations may not have equivalent activity at the same concentration.

Prepare in a manner similar to standard MacConkey's broth, but adjust the pH to 7.4, and distribute in suitable storage containers (e.g. screw-capped bottles) before sterilizing.

(Taylor *et al.*, 1955)

MALT EXTRACT AGAR

(For the culture of yeasts and moulds)

Malt extract	30.0 g
Mycological peptone	5.0g
Agar	15.0g
Distilled water	1 litre

pH5.4

Dissolve the ingredients in the water by steaming. Distribute as required and sterilize by autoclaving at 121 °C for 15min.

To inhibit the growth of bacteria, antibiotics may be added as sterile solutions to the molten medium immediately before pouring the plates, or the medium may be acidified to pH3.5. Acidification may be achieved by adding aseptically sterile 10% lactic acid (or citric acid) solution to the molten medium immediately before pouring the plates. The exact amount of acid to be added will depend on the make or even the batch of the constituents used.

Sucrose (20%) may be added to this medium to make it suitable for osmophilic counts.

MANNITOL EGG-YOLK PHENOL RED POLYMYXIN AGAR

(For the detection and differentiation of *Bacillus cereus)*

Basal medium

Peptone	10.0g
Meat extract	1.0 g
D-Mannitol	10.0g
Sodium chloride	10.0g
Phenol red (0.2% solution)	12.5ml
Agar	15.0g
Distilled water	887.5 ml

pH7.1

Dissolve all the ingredients in the water by steaming. Distribute in 90-ml amounts in screw-capped bottles, and sterilize at 121°C for 15min.

Polymyxin B sulphate solution

Dissolve 50 mg of polymyxin B sulphate in 50ml of distilled water. Sterilize by membrane filtration.

Preparation of complete medium

To 90ml of molten medium, cooled to 45-50°C, add with aseptic precautions 10ml of egg-yolk emulsion and 1 ml of sterile polymyxin B sulphate solution. The final concentration of antibiotic in the medium is thus about per ml of medium. Mix well, and pour plates with about 15ml medium in each plate. Dry for 1 h at 45°C before use in order that the medium will absorb the inoculum liquid during surface colony count procedures.

(Mossel *et.al.*, 1967)

MODIFIED ALKALINE PEPTONE WATER

(For the isolation of *Vibrio)*

Peptone	10.0g
Sodium chloride	10.0 g
Magnesium chloride hexahydrate	4.0 g
Potassium chloride	4.0g
Distilled water	1 litre

pH 8.6 ±0.2

Dissolve the ingredients in the water, mix, and adjust the pH to 8.6 at 25°C. Distribute as required and sterilize at 12TC for 15min.

(Roberts *et. al.*, 1995)

NUTRIENT AGAR

(A general purpose culture medium for bacteria)

Nutrient broth	1 litre
Agar	15.0g

pH 7.2 ±0.2

Dissolve the agar in the nutrient broth by autoclaving at 121°C for 20min. Adjust the pH to 7.2. Filter through paper pulp. Distribute as required and sterilize at 121°C for 20 min.

NUTRIENT BROTH

(A general purpose culture medium for bacteria)

Peptone	10.0g
Lab-Lemco meat extract	10.0g
Sodium chloride	5.0 g
Distilled water	1 litre

pH 7.2 ±0.2

Dissolve the ingredients in the water by steaming. Cool, adjust to pH 7.6, and autoclave at 121°C for 15min. Filter and adjust to pH7.2. Distribute as required, and sterilize at 121°C for 20min.

PEPTONE WATER

(Suitable for the indole test)

Tryptone or tryptose	10.0g
Sodium chloride	5.0 g
Distilled water	1 litre

pH 7.1 ±0.2

Dissolve the peptone and sodium chloride in the water by steaming. Adjust to pH 7.2, and dispense in 5-ml amounts in 150 × 16-mm test-tubes and sterilize by autoclaving at 121°Cfor 15min.

PEPTONE WATER DILUENT

(For enumeration and isolation techniques)

Peptone	1.0 g
Distilled water	1 litre

pH 7.1 ±0.2

Dissolve the peptone in the water, adjust to pH 7.0, dispense as required and sterilize by autoclaving at 121°C for 20 min.

(Straka and Stokes, 1957)

PLATE COUNT AGAR (TRYPTONE GLUCOSE YEAST EXTRACT AGAR)

(A non-selective medium for general viable counts of bacteria in foods)

Tryptone	5.0g
Yeast extract	2.5g
D-Glucose	1.0g
Agar	15.0g
Distilled water	1 litre

pH 7.0 ±0.2

Dissolve the ingredients in the water by steaming. Adjust to pH7.0, dispense in 10-ml or 100-ml amounts in screw-capped bottles, and sterilize by autoclaving at 121°C for 15min.

POLYMYXIN ACRIFLAVIN LITHIUM CHLORIDE CEFTAZIDIME AESCULIN MANNITOL (PALCAM) AGAR

(For the isolation of *Listeria monocytogenes)*

Basal medium

Proteose peptone	23.0g
Yeast extract	3.0g
D-Glucose	0.5 g
Starch	1.0g
Sodium chloride	5.0g
Aesculin	0.8 g
Ammonium iron (III) citrate*	0.5 g
Lithium chloride	15.0g
Phenol red	0.08g
Agar	15.0g
Distilled water	960ml

pH 7.2 ±0.1

* The ammonium iron (III) citrate used should be the more pure green form (around 18% iron content).

POLYMYXIN PYRUVATE EGG-YOLK MANNITOL BROMTHYMOL BLUE AGAR (PPEMBA)

(For the enumeration *of Bacillus cereus* in foods)

Basal medium

Peptone	1.0 g
Mannitol	10.0g
Sodium chloride	2.0 g
Magnesium sulphate, hydrated ($MgSO_4 . 7H_2O$)	0.1 g
Disodium hydrogen phosphate	2.5g
Potassium dihydrogen phosphate	0.25 g
Bromthymol blue	0.12g
Agar	15.0g
Distilled water	1 litre

pH 7.2 ±0.2

Add ingredients to the water, allow to soak for 10min and steam or boil to dissolve. Adjust pH to be 7.2 ± 0.2 at 25°C. Dispense in 90-ml amounts and autoclave at 121°C for 15 min.

Polymyxin B solution

Polymyxin B sulphate	1 million units
Distilled water	10ml

Dissolve the polymyxin B sulphate in the water and sterilize by membrane filtration.

Sodium pyruvate solution

Sodium pyruvate	20.0 g
Distilled water	to 100ml

Dissolve the sodium pyruvate in the water and sterilize by membrane filtration.

Cycloheximide solution

Cycloheximide'H'	0.4 g
Distilled water	100ml

Dissolve the cycloheximide in the water and sterilize by membrane filtration.

Preparation of complete medium

To a bottle of molten basal medium, cooled to 50°C, add with aseptic precautions:

(a) 1.0ml of sterile polymyxin B solution
(b) 1.0ml of sterile cycloheximide solution
(c) 5.0ml of sodium pyruvate solution
(d) 5.0ml of sterile egg-yolk emulsion

Mix well and pour into Petri dishes. The prepared plates of complete medium may be stored in plastic bags in the refrigerator for up to 5 days.

(Holbrook and Anderson, 1980; Corry *et al,* 1995a)

POTATO DEXTROSE AGAR

(For the growth of microfungi)

Potatoes, peeled and diced	200 g
D-Glucose	20 g
Agar	15 g
Distilled water to	1 litre

Boil 200 *g* of peeled, diced potatoes for 1 h in 1 litre of distilled water. Filter, and make up the filtrate to 1 litre. Add the glucose and agar and dissolve by steaming. Distribute in 10-or 100-ml amounts in screw-capped bottles and sterilize by autoclaving at 121°C for 20 min.

Note. Potato carrot dextrose agar can be prepared by substituting 50 g carrot for 50 g potato in the above recipe.

RAPPAPORT-VASSILIADIS ENRICHMENT BROTH

(For the enrichment of *Samonella)*

Solution A

Soya peptone	4.5 g
Sodium chloride (AR grade)	7.2 g
Potassium dihydrogen phosphate	1.44g
Distilled water	1 litre

This must be made on the day of preparation of the complete medium. Mix the ingredients into the water, and heat at 70°C until dissolved.

Solution B

Magnesium chloride hexahydrate ($MgCl_2$-$6H_2O$)	400 g
Distilled water	1 litre

The total volume of this stock solution will be 1260ml, and will contain 31.7g of MgCl2-6H2O per 100ml (Peterz et al, 1989). Magnesium chloride is very hygroscopic, so Vassiliadis (1983) recommends preparing this from a newly opened container of the chemical. The solution may be kept in a dark bottle at ambient temperature for at least 1 year.

Solution C

Malachite green oxalate (analytically pure) 0.4 g Distilled water 100ml

Dissolve the dye in the water. The solution may be stored in a dark bottle for at least 6 months at ambient temperature.

Preparation of complete medium

To 1 litre of Solution A add 100ml of solution B and 10ml of solution C. (The final concentration of $MgCl_2$ -$6H_2O$ in the complete medium will be 28.6 g per litre (Peterz *et al,* 1989). This concentration is critical.)

Distribute as required (e.g. 10-ml amounts in test-tubes, or 500-ml amounts in screw-capped bottles). Sterilize at 115°C for 15min. The medium should be stored refrigerated in the dark and used within 1 month.

This medium is available as a dehydrated medium. However, Peterz *et al.* (1989) found that a number of dehydrated media performed poorly compared with the 'home-made' version. This has been discussed by Busse (1995).

(Vassiliadis, 1983; Peterz *et al,* 1989)

ROGOSA AGAR MODIFIED (MABBITT AND ZIELINSKA)

(For the isolation and enumeration of lactobacilli in fermented milk products)

Milk digest

Mix 1 litre of separated milk or reconstituted milk, adjusted to pH8.5, 5g of trypsin'H', and 10ml of chloroform'111. Incubate at 37°C for 24 h, then steam for 20min and filter while hot. Adjust the pH to 6.65 ± 0.02 with glacial acetic acid'H] (about 0.5 ml per litre) using a pH meter.

Nutrient solution

Yeast extract	12.0g
Di-ammonium hydrogen citrate	4.8 g
Potassium dihydrogen phosphate	14.4g
D-Glucose	48.0 g
Tween 80	2.4 g
Salts solution§	12.0ml
Distilled water	200ml

§Salts solution

Magnesium sulphate, hydrated ($MgSO_4$ $7H_2O$)	11.5 g
Manganese sulphate, hydrated ($MnSO_4 . 4H_2O$)	2.8 g
Ferrous sulphate, hydrated ($FeSO_4 . 7H_2O$)	0.08 g
Distilled water	100ml

Dissolve ingredients in water by gentle heating. Add 60 ml of 4 M sodium acetate-acetic acid buffer at pH 5.37 ±0.02. Make up the volume to 200ml with distilled water.

The salts solution has a final pH of 5.0 and is not sterilized; it should be stored in the refrigerator until required.

Preparation of complete medium

Dissolve 19 g agar in 700ml of milk digest at 121°C for 20min, and while hot mix with 185ml of nutrient solution previously warmed to 50°C. Make up the volume to 1 litre with hot digest.

Distribute the medium aseptically in 10-ml amounts into sterile screw-capped bottles, and store in the refrigerator (without sterilizing) until required.

The medium should be prepared for use with as little heating as possible to avoid darkening and the formation of a precipitate.

(Mabbitt and Zielinska, 1956)

Appendices

APPENDIX I

SAFETY PERCAUTIONS IN MICROBIOLOGICAL LABORATORY

In the Microbiological laboratory there are medias, chemicals, flamable liquids, and infectious materials. Mishanding in the laboratory can be dangerous not only to the individual at fault but to others who are working nearby.

The analysis of foods for pathogens and toxins might involve risk and may be a source of slow poisioning or of disease.

The main routes of entery of infections to the body are inhalation by ingestions to the body, through cuts and abrasions and infecting the eyes. By intelligent using the Medias and chemicals it has been established that such risks have been minimized almost to the point to extinction.

Some important precautions to be taken during working in the laboratory

1. Do not eat, drink and smoke in the laboratory premisis.
2. Wear a laboratory coat.
3. Used laboratory. Coats should be autoclaved before being sent to laundering.
4. Wear disposable surgical gloves when handling infectious or toxic materials, those are autoclaved before throughing in the dustbin.
5. Wear goggle and mask when handling botulinium and weighing Dehydrated Media.

6. Do not touch pencil and labels with tongue.
7. After each work phase, wipe the bench thoroughly with disinfectant and wash your hands.
8. Plug top ends of pipettes with cotton before sterilization.
9. Any spilled culture and left out food should immediatly be removed and the surface should be cleaned with disinfectant.
10. Always keep first-Aid-Box for any accident.

APPENDIX II

The values obtained form the tables are based of assumption that the samples adequatly mixed and correctly duluted. The microorganisms are properly distributed at the time of making dilutions.

These assumptions may sometimes be incorrect. In an experiment, when no growth occurs in lower dilutions and growth occurs in higher dilutions it could be indicative of an antimicrobial inhibitory substance present in the food which becomes diluted below a threshold minimum inhibitory concentration before the microorganisms have been diluted out to undetictable low numbers. The 1st, 2nd dilutions show no growth. In the tables B and C positive and negative tubes are only listed which are likely to occur when anydoubt occurs after seeing the sample history, the experiment is repeated and the result should not be used as the basis of quality assurance decesion.

PROBABILITY TABLES FOR THE ESTIMATION OF MICROBIAL NUMBERS

Table A : Values of the MPN for two tubes inoculated from each of these successive tenfold dilutions.

No. of positive tubes observed At each dilution			
Ist dilution	*2nd dilution*	*3rd dilution*	*MPN of microorganisms per inoculum of the first dilution*
0	0	0	0
0	0	1	0.45
0	1	0	0.46
1	0	0	0.6
1	0	1	1.2
1	1	0	1.3
1	1	1	2.0
1	2	0	2.1
2	0	0	2.3
2	0	1	5.0
2	1	0	6.2
2	1	1	13
2	1	2	21
2	2	0	24
2	2	1	70
2	2	2	100+

Approximate 95% confidence limits may be calculated (Cochran 1950)as:

$$\frac{\text{MPN}}{6.61} \text{ to MPN} \times 6.61$$

Table B : Values of the MPN for three tubes inoculated from each of three successive tenfold dilutions (after Demeter et al., 1933)

No. of positive tubes observed at each dilution			*MPN of microorganisms per inoculum of 1st dilution*	*Category**	*95% C.I.†*
1st dilution	*2nd dilution*	*3rd dilution*			
0	0	0	0	—	
0	0	1	0.3	3	
0	1	0	0.3	2	<0.1–1.7
0	1	1	0.6	4	
0	2	0	0.6	4	
1	0	0	0.4	1	<0.1–2.1
1	0	1	0.7	3	
1	0	2	1.1	4	
1	1	0	0.7	2	0.2–2.8
1	1	1	1.1	4	
1	2	0	1:1	3	
1	2	I	1.5	4	
1	3	0	1.6	4	
2	0	0	0-9	1	0.2 3.8
2	0	1	1.4	3	
2	0	2	2.0	4	
2	1	0	1.5	2	0.5 -5.0
2	1	1	2.0	4	
2	1	2	3.0	4	
2	2	0	2.0	3	
2	2	1	3.0	4	
2	2	2	3.5	4	
2	2	3	4.0	4	
2	3	0	3.0	4	
2	3	1	3.5	4	
2	3	2	4.0	4	
3	0	0	2.5	1	< 1–13
3	0	1	4.0	2	1–18
3	0	2	6.5	4	
3	1	0	4.5	1	1–2 1

No. of positive tubes observed at each dilution			MPN of microorganisms per inoculum of 1st dilution	Category*	95% C.I.†
1st dilution	2nd dilution	3rd dilution			
3	1	1	7.5	2	2–28
3	1	2	11.5	3	
3	1	3	16.0	4	
3	2	0	9.5	1	3 38
3	2	1	15.0	2	5–50
3	2	2	20.0	3	
3	2	3	30.0	4	
3	3	0	25.0	1	<10–140
3	3	1	45.0	1	10–240
3	3	2	110.0	1	30–480
3	3	3	140.0 +	-	

* In the long run, category 1 combinations may be expected to constitute 67.5% of test results containing both postive and negative tubes; categories (1 +2) 91 %; and categories (1+2 + 3) 99% of such test results. Category 4 and unlisted combinations are highly unlikely (Woodward, 1957). Combinations from categories 3 and 4 should not be used as the basis for quality assurance decisions involving the rejection and/or reprocessing of batches of food: samples should be retested. In the event of large numbers of improbable combinations being obtained, experimental procedures should be examined closely, as this is indicative of the presence of disturbing influences (e.g. improper mixing of samples and/or dilutions, presence of antagonistic substances in the foodstuff, etc.).

§ The 95% confidence intervals (as calculated by de Man (1975)) are given for category 1 and 2 results. Since it is likely that the experimental conditions for a microbiological analysis resulting in a category 3 or 4 combination will not have been in statistical control (i.e. the assumptions described in the introduction to this Appendix are incorrect), 95% confidence intervals are not given for these combinations.

Table C : Values of the MPN for five tubes inoculated from each of three successive tenfold dilutions (after Woodward 1957; Report, 1969)

No. of positive tubes observed at each dilution			*MPN of microorganisms per inoculum of 1st dilution*	*Category**	*95% C.I.†*
1st dilution	*2nd dilution*	*3rd dilution*			
0	0	0	0		
0	0	1	0.2	3	
0	0	2	0.4	4	
0	1	0	0.2	3	
0	1	1	0.4	4	
0	1	2	0.6	4	
0	2	0	0.4	3	
0	2	1	0.6	4	
0	3	0	0.6	4	
1	0	0	0.2	1	<0.1-1.1
1	0	1	0.4	3	
1	0	2	0.6	4	
1	1	0	0.4	2	0.1-1.5
1	1	1	0.6	4	
1	1	2	0.8	4	
1	2	0	0.6	3	
1	2	1	0.8	4	
1	3	0	0.8	4	
1	3	1	1.0	4	
2	0	0	0.4	1	0.1-1.7
2	0	1	0.7	3	
2	0	2	0.9	4	
2	0	3	1.2	4	
2	1	0	0.7	2	0.2-2.1
2	1	1	0.9	3	
2	1	2	1.2	4	
2	2	0	0.9	3	
2	2	1	1.2	4	
2	2	2	1.4	4	
2	3	0	1.2	4	

No. of positive tubes observed at each dilution			*MPN of microorganisms per inoculum of 1st dilution*	*Category**	*95% C.I.†*
1st dilution	*2nd dilution*	*3rd dilution*			
2	3	1	1.4	4	
3	0	0	0.8	1	0.3-2.4
3	0	1	1.1	3	
3	0	2	1.3	4	
3	1	0	1.1	2	0.4-2.9
3	1	1	1.4	3	
3	1	2	1.7	4	
3	1	3	2.0	4	
3	2	0	1.4	3	
3	2	1	1.7	4	
3	2	2	2.0	4	
3	3	0	1.7	3	
3	3	1	2.1	4	
3	4	0	2.1	4	
3	4	1	2.4	4	
4	0	0	1.3	1	0.5-3.8
4	0	1	1.7	3	
4	0	2	2.1	4	

Table D

No. of positive tubes observed at each dilution			*MPN of microorganisms per inoculum of 1st dilution*	*Category**	*95% C.I.†*
1st dilution	*2nd dilution*	*3rd dilution*			
4	0	3	2.5	4	
4	1	0	1.7	2	0.7-4.6
4	1	1	2.1	3	
4	1	2	2.6	4	
4	2	0	2.2	2	0.9-5.6
4	2	1	2.6	3	
4	2	2	3.2	4	
4	3	0	2.7	3	
4	3	1	3.3	3	
4	3	2	3.9	4	
4	4	0	3.4	3	
4	4	1	4.0	4	
4	4	2	4.7	4	
4	5	0	4.1	4	
4	5	1	4.8	4	
4	5	2	5.6	4	
5	0	0	2.3	1	0.9-8.6
5	0	1	3.1	3	
5	0	2	4.3	3	
5	0	3	5.8	4	
5	1	0	3.3	1	1-12
5	1	1	4.6	2	2-15
5	1	2	6.4	4	
5	1	3	8.4	4	
5	1	4	11.5	4	
5	2	0	4.9	1	2-17
5	2	1	7.0	2	3-21
5	2	2	9.5	3	
5	2	3	12.0	4	
5	2	4	15.0	4	
5	3	0	7.9	1	3-25
5	3	1	11.0	2	4-30

No. of positive tubes observed at each dilution			*MPN of microorganisms per inoculum of 1st dilution*	*Category**	*95% C.I.†*
1st dilution	*2nd dilution*	*3rd dilution*			
5	3	2	14.0	3	
5	3	3	17.5	3	
5	3	4	20.0	4	
5	3	5	25.0	4	
5	4	0	13.0	1	5-39
5	4	1	17.0	2	7-48
5	4	2	22.0	2	10-58
5	4	3	28.0	3	
5	4	4	35.0	3	
5	4	5	42.5	4	
5	5	0	24.0	1	10-94
5	5	1	35.0	1	10-130
5	5	2	54.0	1	20-200
5	5	3	92.0	1	30-290
5	5	4	160.0	1	60-530
5	5	5	180.0 +	-	

* See footnote to Table B.

† See footnote to Table B.